Advanced Model Order Reduction Techniques in VLSI Design

Model order reduction (MOR) techniques are important in reducing the complexity of nanometer VLSI designs, and consequently controlling "parasitic" electromagnetic effects, so that higher operating speeds and smaller feature sizes can be achieved. This book presents a systematic introduction to, and treatment of, the key MOR methods used in general linear circuits, using real-world examples to illustrate the advantages and disadvantages of each algorithm.

Starting with a review of traditional projection-based techniques and proofs of some fundamental theories, coverage progresses to advanced "state-of-the-art" MOR methods for VLSI design. These include HMOR, passive truncated balanced realization (TBR) methods, efficient inductance modeling via the VPEC model, general model optimization and passivity enforcement methods, passive model realization techniques, and structure-preserving MOR techniques. Numerical methods have been used throughout and, where possible, approached from the CAD engineer's perspective. This avoids complex mathematics, and allows the reader to take on real design problems and develop more effective tools.

With practical examples and over 100 illustrations, this book is suitable for researchers and graduate students of electrical and computer engineering, as well as for practitioners working in the VLSI design and design automation industries.

SHELDON X.-D. TAN is an associate professor in the Department of Electrical Engineering, and cooperative faculty member in the Department of Computer Science and Engineering, at the University of California, Riverside. He received his Ph.D. in electrical and computer engineering in 1999 from the University of Iowa, Iowa City. His current research interests focus on design automation for VLSI integrated circuits.

LEI HE is an associate professor in the Department of Electrical Engineering at the University of California, Los Angeles, where he was also awarded his Ph.D. in computer science in 1999. His current research interests include computer-aided design of VLSI circuits and systems.

Advanced Model Order Reduction Techniques in VLSI Design

SHELDON X.-D. TAN
University of California, Riverside

LEI HE
University of California, Los Angeles

CAMBRIDGE
UNIVERSITY PRESS

CAMBRIDGE UNIVERSITY PRESS
Cambridge, New York, Melbourne, Madrid, Cape Town, Singapore, São Paulo

Cambridge University Press
The Edinburgh Building, Cambridge CB2 8RU, UK

Published in the United States of America by Cambridge University Press, New York

www.cambridge.org
Information on this title: www.cambridge.org/9780521865814

First published 2007

Printed in the United Kingdom at the University Press, Cambridge

A catalog record for this publication is available from the British Library

ISBN 978-0-521-86581-4 hardback

Contents

Figures

Tables

Foreword

Interconnect model reduction has emerged as one crucial operation for circuit analysis in the last decade as a result of the phenomenon of interconnect dominance of advanced VLSI technologies. Because interconnect contributes to a significant portion of the system performance, we have to take into account the coupling effects between subcircuit modules. However, the extraction of the coupling renders many small fragments of parasitics. While the values of the parasitics are small, the number of fragments is huge and this makes the accumulated effect non-negligible. If left untreated, the amount of parasitics can gobble up the memory capacity and consume long CPU time during circuit analysis.

Model reduction transforms a system into a circuit of much smaller size to approximate the behavior of the original description. Many researchers have contributed to the advancement of the techniques and demonstrated drastic reduction of the circuit sizes with satisfactory output responses in published reports. Many of these techniques have also been implemented in software tools for applications. However, it is important for the users to understand the techniques in order to use the package properly. To adopt these approaches, we need to inspect the following features.

1. Efficiency of the reduction: the complexity of the reduction algorithm determines the CPU time of the model reduction. The size of the reduced circuit affects the simulation time.

2. Reduction of both model order and terminals of circuits: reduction of terminals was investigated less in the past and combined terminal and model order reduction leads to more compact models.

3. Robustness of the algorithms: the numerical stability of the reduction algorithm ensures the robustness of the operation.

4. Structure of the reduced systems: the reduced systems may or may not preserve important characteristics like symmetry, reciprocity, etc. Those structure characteristics are important for reduction itself and for systems using the models.

5. Realizablility of the reduced system: the reduced system is realizable if it is passive and we can implement it using electrical elements with positive or negative values. We can simulate a realizable system with general simulation tools. Otherwise, we need to check if the reduced system satisfies the constraints of the simulation package.

6. Passivity of the reduced circuits: the passivity ensures that the simulation

outputs are bounded for bounded inputs even if the reduced circuit is combined with other passive subcircuits.

7. Error bounds: The error bounds of the output responses provide users with confidence in the results.

In this book, Professors Sheldon X.-D. Tan and Lei He presented a comprehensive description of the reduction techniques. They have provided motivations for the approaches and insights into the algorithms as active researchers in the field. I found that the treatment of the subject is innovative and the general description is pleasant to read. The book covers the contemporary results and opens windows on future research directions in the field.

Chung-Kuan Cheng
Department of Computer Science and Engineering,
The University of Californiat at San Diego

Acknowledgments

Sheldon Tan is deeply indebted to the support, encouragement and love of his family: his wife, Yan Ye, his lovely daughters Felicia Tan and Lesley Tan. Sheldon Tan is also grateful to his parents, Linsen Tan, Wanhua Mao, for constant encouragement, and inspiration over the years.

Many people contributed to this book. The authors would like thank some of their students, Boyuan Yan, Pu Liu, and Duo Li from University of California at Riverside for their dedicated work and the implementation of some of algorithms discussed in this book. Boyuan Yan also made significant contributions to Chapters 3 and 4 and Pu Liu made noticeable contributions to Chapter 6. We will thank Dr. Hao Yu from UCLA for his contribuation to Chapters 7 and 8.

Some of the materials in this book come from the graduate level course notes of EE213, Computer-aided electronic circuit simulation, taught by Sheldon Tan at UC Riverside, who would like to thank his students in this course for developing those notes in the past three years.

This book would never have been completed without support from many other sources. Sheldon Tan is grateful to Dr. Zhihong Liu, Dr. Lifeng Wu, Dr. Bruce McGaughy from Cadence Design Systems Inc. for their strong supports and collaboration for model order reduction related projects at University of California at Riverside.

Sheldon Tan would like to acknowledge the support of the National Science Foundation CAREER Award under grant No. CCF-0448534, NSF awards under grant No. CCF- 0541456 and No. OISE-0623038. University of California MICRO program under grant No. 04-088, No. 05-111 and No. 06-252 via Cadence Design Systems and University of California at Riverside through UC Regent's Faculty Fellowship, UC Senate Research Funds. These supports are essential for development of the book.

Special thanks also to Ms. Anna Littlewood, Ms. Jeanette C. Alfoldi and Dr. Philip Meyler at Cambridge University Press for soliciting the proposal for this book and for their dedicated help, hard work and encouragement throughout the course of writing this book.

1 Introduction

1.1 The need for compact modeling of interconnects

As VLSI technology advances into the sub-100nm regime with increased operating frequency and decreased feature sizes, the nature of the VLSI design has changed significantly. One fundamental paradigm change is that parasitic interconnect effects dominate both the chip's performance and the design's complexity growth. As feature sizes become smaller, their electromagnetic couplings become more pronounced. As a result, their adverse impacts on circuit performances and powers will become more significant. Signal integrity, crosstalk, skin effects, substrate loss and digital and analog substrate couplings are now adding severe complications to design methodologies already stressed by increasing device counts. It was observed that today's high performance digital design essentially becomes analog circuit design [24] as there has been a need to observe a finer level of detail.

In addition to dominant deep submicron effects, the exponential increase of device counts causes a move in the opposite direction: we need to increase the increasing design abstraction levels to cope with the design capacity growth. It was widely believed that behavioral and compact modeling for the purpose of synthesis, optimization, and verification of the complicated system-on-a-chip are viable solutions to address these challenging design problems [66].

In this book, we focus on the compact modeling of on-chip interconnects and general linear time invariant systems (LTI) because interconnect parasitics, which are modeled as linear RLCM circuits[1.1], are the dominant factors for complexity growth. Unchecked parasitics from on-chip interconnects and off-chip packaging will de-tune the performance of high-speed circuits in terms of slew rate, phase margin and bandwidth [2]. Reduction of design complexity especially for the extracted high-order RLCM networks is crucial for reducing the explosive design productivity gap in the nanometer VLSI design and verification.

This book does not by any means intend to be comprehensive. The absence of coverage of work by other researchers should not diminish their contributions.

1.1 *M here means the mutual inductances.*

1.2 Interconnect analysis and modeling methods in a nutshell

Compact modeling of passive RLC interconnect networks has been a research intensive area in the past decade due to increasing adverse deep submicron effects and interconnect-dominant delays in current high-performance VLSI designs [23, 72]. A number of projection-based model order reduction (MOR) based techniques have been introduced [32, 33, 38, 85, 91, 113, 114] to analyze the transient behavior of interconnects.

An asymptotic waveform evaluation (AWE) algorithm was first proposed [91, 92] where explicit moment matching was used to compute the dominant poles via Pade approximation. The AWE algorithm used the moment concept to control and measure the accuracy of the reduced and the original system, and this was advantageous over many previous black-box fitting methods. Also the AWE method shows that the wildly popular interconnect delay model, The Elmore delay model, is just the first order of moments of a circuit [28]. The success of the AWE method led to intensive research efforts on model order reduction of interconnect circuits.

The AWE method is numerically unstable for higher-order moment approximation. The approximation is carried out around $s = 0$. The same authors introduced some remedial methods to overcome this problem by frequency shifting and expanding around $s = \infty$ [92], but a more effective method involved carrying out multiple-point expansions along the imaginary axis (called frequency hopping) and combining the expansion results at higher computing costs [19].

A more elegant solution to the numerical problem of AWE is to use projection-based model order reduction (MOR) methods, which are based on implicit moment matching. The main idea is to project the explicit moment space into an orthonormal subspace, called Krylov subspace. The projection process basically preserves the moment information, but the Krylov vectors contain much less numerical noise compared with the explicit moments owing to the generation of Krylov subspace. The Pade via Lanczos (PVL) method was the first projection-based method [32], where the Lanczos process, which is a numerically stable method for computing eigenvalues of a matrix, was used to compute the Krylov subspace. Feldmann also proved that the reduced system implicitly matches the original system to a certain order of moments. Later on, the PVL method was extended to deal with multiple input and multiple output cases by MPVL [33], and to deal with circuits with symmetric matrices by SyPVL algorithm [38]. The Krylov subspace can also be generated by the Arnoldi process, which is based on the so-called orthogonal projection. Examples include the Arnoldi method [114] and Arnoldi transformation method [113]. But Arnoldi methods only match the half order of the moments or block moments for the same reduced order.

To ensure the passivity of the reduced model further, it was shown in [63] that congruence transformation can preserve the model's passivity if the system matrices are in a *passive form*. Later PRIMA [85] used the Krylov subspace vectors to form the projector for the congruence transformation, which leads to passive models with the matched moments in the rational approximation paradigm. Projection-based methods, however, have several drawbacks. First, they are not efficient for circuits with many inputs and output terminals. This reflect in the fact that the reduction cost is tied to the number of terminals; the number of poles of reduced models is also proportional to the number of terminals. Second, PRIMA-like methods do not preserve structure properties like reciprocity of a network. Third, it is difficult to apply PRIMA-like methods to model very high frequency circuits where the circuit parameters are frequency dependent or where only measure data are available in terms of scattering-parameters.

Another approach to circuit-complexity reduction is by means of local node reduction. The main idea is to reduce the number of nodes in the circuits and approximate the newly added elements in the circuit matrix in reduced rational forms. The major advantage of these methods over projection-based methods is that the reduction can be made in a local manner and no overall solutions of the whole circuit are required (with some circuit realization or synthesis techniques), which makes those methods very amenable to attacking large linear networks. This idea has been explored by approximate Gaussian elimination for RC circuits [27], by the TICER program [107], which is also based on Gaussian elimination but only keeps first two moments, and by the extension of TICER method into RLC circuits [3]. The rational approximation is also explored by the direct truncation of the transfer function algorithm (DTT) [56] for tree-structured RLC circuits and by an extended DTT method for non-tree structured RLC circuits [125].

Recently, a more general topology-based node-reduction method was proposed [96, 98], in which nodes are reduced one at a time (topologically, it is called Y-Δ transformation) and the generated admittance in the reduced network is represented as an order-reduced rational function of s. This method is equivalent to symbolic Gaussian elimination (s is the only symbol) but the reduction is made on circuit topologies only, which is equivalent to the nodal analysis (NA) formulation of a circuit only. The stability is enforced by Hurwitz polynomial approximation. But this method only works for linear circuits with limited element types (RCLK-VJ) and cannot be applied to reduce general linear circuits due to NA formulation requirement. A more general multi-node or block version of Y-Δ transformation, named *hierarchical model order reduction or HMOR* was proposed by Tan [121,124]. Since a number of nodes can be reduced at the same time, this method essentially leads to the general node-reduction based hierarchical model order reduction.

The third major development for model order reduction of LTI systems is by means of control-theoretical-based truncated balance realization (TBR) methods, where the weak uncontrollable and unobservable state variables are truncated to achieve the reduced models [81, 87, 89, 131]. The TBR methods can produce nearly optimal models but they are more computationally expensive than projection-based methods. Also, TBR can produce the passive models by so called *positive real TBR methods* [87, 131]. Recently, empirical TBR method, named poor man's TBR, was proposed to improve the scalability of the TBR methods, which shows the connection with the generalized projection-based reduction methods [89].

1.3 Book outline

Mode order reduction of time-invariant linear systems is still an active research area. There are many excellent books covering the classic MOR methods such as moment matching, Krylov subspace projection-based methods and node-reduction based methods [14, 20, 72, 97].

In this book, we look at some important developments in this area since the PRIMA method was introduced in 1997, but we make no attempt to be comprehensive. Instead, we primarily present several important methods that give new perspectives on model order reduction techniques in terms of improved efficiency, accuracy, and more compact reduced model sizes over the existing projection-based methods. For instance, we look at truncated balanced realization-based methods, the hierarchical model order reduction method, MOR methods for linear circuits with multiple terminals, MOR methods for highly inductive circuits, general passivity enforcement and circuit realization techniques, and terminal reduction methods. In the following, we give the outline of this book.

- Chapter 2 will review the concepts of model order reduction, moment matching, and classic explicit moment-matching methods like AWE for model reduction. Then we will review Krylov subspace projection-based model order reduction techniques, such as the projection-based MOR methods, which are still the most widely used reduction techniques. We will present the basic concepts of Krylov subspace, passivity, numerical algorithms, such as Arnoldi and Lanczos methods for obtaining orthogonal Krylov basis and reduction matrices, and the PRIMA method. We also present some important theoretical results regarding the Krylov subspaces. Some of the concepts introduced here will be used throughout this book.
- Chapter 3 studies the SVD-based model order reduction technique based on the classic control theory, called truncated balanced realization (TBR), which leads to more compact models than the Krylov subspace projection MOR methods but at much higher computation costs. We will review the basic concepts of truncated balanced realization methods in terms of controllability and observability from the control-theory perspective. We then present the positive real TBR methods which can produce the passive models. After this, the empirical TBR method named poor man's TBR is also presented, which can scale to reduce large circuits. Finally, some numerical and implementation issues with TBR methods are discussed.
- Chapter 4 presents a new passive TBR method, called *PriTBR*, for interconnect modeling. Different from existing passive truncated balanced realization (TBR) methods where numerically expensive Lur'e or algebraic Riccati equations (ARE's) are solved, the new method performs balanced truncation on the linear systems in descriptor form by solving generalized Lyapunov equations. Passivity preservation is achieved by congruence transformation instead of simple truncations. The PriTBR method can be applied as a second stage model

order reduction to work with Krylov subspace methods to generate a nearly optimal reduced model from a large scale interconnect circuit while passivity, structure, and reciprocity are preserved at the same time.

- In Chapter 5, we present the hierarchical model order reduction method, named *HMOR*, which is based on multiple-point expansion. We will review basic steps of the hierarchical reduction technique and describe the flow of the multiple-point expansion based on hierarchical reduction. The concept of symbolic analysis based on a determinant decision diagram (DDD) will be reviewed; this is the core algorithm for the HMOR. We also discus the new pole search algorithm and some important properties of hierarchical reductions such as structure preserving and numerical stability for tree-like circuits.
- Chapter 6 first reviews a terminal reduction algorithm named *SVDMOR*, which performs the reduction on the input and output position matrices of a transfer function matrix. Then we present another general terminal reduction algorithm, named *TermMerg*, to efficiently reduce the terminal number of general linear interconnect circuits with a large number of input and output terminals considering delay variations. TermMerg can reduce many similar terminals and keep a small number of representative terminals. It can also work with passive model reduction algorithms to generate passive compact models. This is in contrast to SVDMOR, which may not produce passive models. After terminal reduction, traditional model order reduction methods can be applied and achieve more compact models and improve simulation efficiency.
- Chapter 7 deals with a new inductance modeling technique, vector potential equivalent circuit and its application in the HMOR. We will discuss the concept of VPEC models and VPEC-based model mutual inductance sparsification technique. Some theoretical results of passivity of VPEC models and its application in the hierarchical modeling reduction will be presented.
- Chapter 8 presents a structure-preserving projection-based MOR method. It starts with the SPRIM method, which is the first structure-preserving reduction algorithm based on 2×2 partitioning of circuit matrices. Then we present a general block structure-preserving projection-based model order reduction technique, called *TBS*, which is an extension of the SPRIM based algorithm. The new algorithm can preserve the structure of the reduced circuits, which makes it easier and more efficient to realize the reduced circuits. Also, we show that by partitioning the original circuits into many disjoint subcircuits, not only can we preserve sparsity of the reduced circuits, but we can also match more poles of the original systems, thus improving the model accuracy.
- Chapter 9 introduces another generalized block structure-preserving reduced order interconnect macromodeling method (BSPRIM). The new approach extends the structure-preserving model order reduction (MOR) method SPRIM [37] into more general block forms. The chapter first shows how a SPRIM-like structure-preserving MOR method can be extended to deal with admittance RLC circuit matrices and show that the $2q$ moments are still matched and symmetry is preserved. It then shows that $2q$ moment match-

ing can't be achieved when the RLC circuits are driven by both current and voltage sources. Using BSPRIM improves SPRIM by introducing the re-orthonormalization process on the partitioned projection matrix. The BSPRIM method can deal with more circuit partitions and can perform the general block structure preserving MOR for circuits formulated in impedance and admittance forms. The reduced models by the proposed BSPRIM will still match the $2q$ moments and preserve the circuit structure properties, like symmetry, as SPRIM does.

- Chapter 10 examines some effective methods to enforce the passivity of models and optimize models. Model passivity and passivity enforcement are widely used for building realizable models from direct measurements and simulation data for radio-frequency and microwave applications. They are also used in HMOR and TermMOR methods. The studied methods include convex-based passivity enforcement and optimization and least-square-based methods for active model optimization.
- Chapter 11 studies the problem of realizing a reduced model into a SPICE-compatible netlist. This process is called model realization. We first present a traditional one-port network synthesis technique, Brune's method, for realizing a passive circuit from its mathematical model. The concept of traditional network realization will be covered. We then present a general multi-port network-based realization technique, which includes a one-port realization based on a Foster's method and general multiple-port impedance realization based on one-port realization techniques. We also discuss how to realize general non-symmetric circuits.
- Chapter 12 presents a novel compact reduced modeling technique to reduce interconnect circuits with many external ports called *TermMOR*. The proposed method overcomes the difficulty associated with subspace projection-based MOR methods for reducing circuits with many ports. The new method can lead to much smaller reduced models for a given frequency range or much higher accuracy given the same model sizes than subspace projection-based methods. Like HMOR, the TermMOR method is a closed-loop method, as it can produce models matching the desired frequency range precisely.
- Chapter 13 presents an approach to enforcing the passivity of a reduced system of general passive linear time-invariant circuits. Instead of making the reduced models passive for infinite frequencies, the method works on the signal waveform driving reduced models. It slightly shapes the waveforms of the signal such that the resulting signal spectra are band limited to the frequency range in which the reduced system is passive. As a result, the reduced models only need to be band-limited passive (also called conditionally passive), which can be achieved much more easily than traditional passivity for a reduced system, especially for one with many terminals or requiring wide band accuracy (more poles). We propose to use spectrum truncation via FFT and IFFT and low-pass-filter-based approaches for transient waveform shaping processing. We analyze the delay and distortion effects caused by using low-pass filters and present methods

to mitigate the two effects.

1.4 Summary

In this chapter, we first present the cases for compact modeling for interconnect circuits. We then briefly survey the previous developments on this topic and present what will be covered in this book for each chapter.

Throughout the book, numerical examples are provided to shed light on the discussed topics to help the reader gain more insights into the discussed algorithms. All our treatments of many topics may not be mathematically rigorous. Instead, we try to present the topics from a typical computer-aided design (CAD) engineer's perspective and try to help reader to apply those techniques to solve real VLSI design problems and develop more efficient simulation tools.

2 Projection-based model order reduction algorithms

Compact modeling of passive RLC interconnect networks has been an intensive research area in the past decade owing to increasing signal integrity effects and interconnect-dominant delay in current system-on-a-chip (SoC) design [72].

In this chapter, we briefly review the existing modeling order reduction (MOR) algorithms for linear time-invariance (LTI) systems developed over the past two decades in the electrical computer-aided design community. Since compact modeling of TLI systems is a well researched and studied field, many efficient approaches have been proposed over the years. Given the space in this book, we cannot review all of them and neither do we attempt to be complete in our review. Instead, we mainly review the Krylov subspace projection-based model order reduction methods, which are widely used MOR methods and are closely related to the rest of this book. Although there exists an excellent and detailed treatment of Krylov subspace projection-based methods already [14], for the completeness of this book, we still present some basic concepts, algorithms and important results for Krylov subspace projection-based MOR methods. We try to present them in a way that can be easily understood from the practical application point of view.

2.1 Moments and moment-matching methods

In this section, we briefly review the concepts of time-domain moments, the Elmore delay and Pade-approximation-based moment-matching method, which are important concepts for subspace projection-based model order reduction methods.

2.1.1 *Concept of moments*

In the s domain, the transfer function of a linear network $H(s)$ is defined as the ratio of the output to the input under zero initial conditions:

$$H(s) = \frac{Y(s)}{X(s)}. \tag{2.1}$$

If the input is the impulse function $\delta(t)$, its Laplace transformation is 1. So the transfer function is also the impulse response at the port. If we expand $H(s)$ around

$s = 0$ by the Taylor series expansion, we have

$$H(s) = \sum_{k=0}^{\infty} m_k s^k, \tag{2.2}$$

where

$$m_k = \frac{1}{k!} \times \left. \frac{\mathrm{d}^k H(s)}{\mathrm{d}s^k} \right|_{s=0}. \tag{2.3}$$

where the kth coefficeint of $H(s)$, m_k, is called the kth moment.

Assuming that $h(t)$ is the corresponding time-domain impulse response, we have

$$H(s) = \int_0^{\infty} \mathrm{e}^{-st} h(t) \mathrm{d}t. \tag{2.4}$$

We rewrite moments defined in (2.4) in terms of $H(t)$ by using the Taylor expansion of e^{-st} in the Laplace transform $H(s)$ and we have

$$\begin{aligned} H(s) &= \int_0^{\infty} h(t) \mathrm{e}^{-st} \mathrm{d}t \\ &= \int_0^{\infty} \left(1 - st + s^2 \frac{t^2}{2} + \cdots + s^k \frac{(-1)^k}{k!} t^k + \cdots \right) \mathrm{d}t \\ &= \sum_{k=0}^{\infty} s^k \frac{(-1)^k}{k!} \int_0^{\infty} t^k h(t) \mathrm{d}t. \end{aligned} \tag{2.5}$$

Comparing (2.5) with the definition that $H(s) = \sum_{k=0}^{\infty} m_k s^k$, moments can be rewritten as:

$$m_k = \frac{(-1)^k}{k!} \int_0^{\infty} t_k H(t) \mathrm{d}t, \tag{2.6}$$

or

$$m_0 = \int_0^{\infty} h(t) \mathrm{d}t, \tag{2.7}$$

$$m_1 = - \int_0^{\infty} t h(t) \mathrm{d}t, \tag{2.8}$$

$$m_2 = \frac{1}{2!} \int_0^{\infty} t^2 h(t) \mathrm{d}t, \tag{2.9}$$

$$\ldots$$

2.1.2 *Elmore delay*

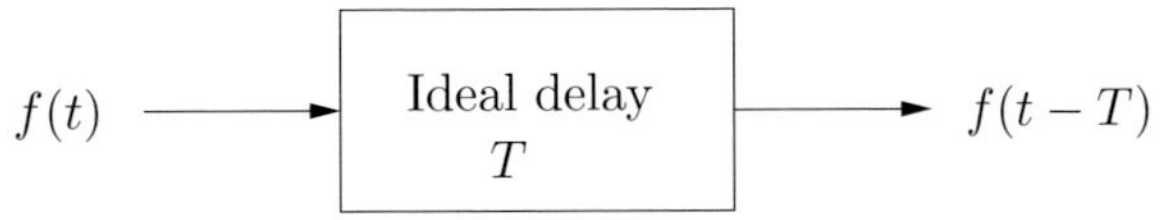

Figure 2.1 The network of an ideal delay of T.

For an ideal delay network as shown in Figure 2.1, the response of the network for an input function $f(t)$ is $f(t-T)$. We take the Laplace transformation of $f(t-T)$ and we have

$$\mathcal{L}(f(t-T)) = \mathrm{e}^{-sT}F(s), \tag{2.10}$$

where $F(s)$ is the Laplace transformation of $f(t)$ and $\mathcal{L}(*)$ is the Laplace transformation operator. So the ideal delay element's transfer function is

$$H_d(s) = \mathrm{e}^{-sT}. \tag{2.11}$$

If we take the derivative of $H_d(s)$ with respect to s, we have

$$\frac{\mathrm{d}H_d(s)}{ds}|_{s=0} = -T. \tag{2.12}$$

As a result, we may use $\frac{\mathrm{d}H_d(s)}{ds}|_{s=0}$ as an approximate for the delay of a general linear network described by $H(s)$, i.e.,

$$T_d \approx -\frac{\mathrm{d}H_d(s)}{ds}|_{s=0}. \tag{2.13}$$

T_d is the so-called Elmore delay [28], which is also the first-order moment in (2.3). So we have

$$m_1 = \int_0^\infty h(t)t\mathrm{d}t = \frac{\mathrm{d}H_d(s)}{ds}|_{s=0}. \tag{2.14}$$

Another popular mathematic interpretation of the Elmore delay is by means of probability perspective. Physically, the delay of a network can be measured using the 50% point delay of the monotonic step response from a unit step input. If $h(t)$ is the unit impulse response, the unit step response is $\int_o^\infty h(t)\mathrm{d}t$. The 50% point delay τ is then defined as

$$\int_0^T h(t)\mathrm{d}t = 0.5. \tag{2.15}$$

If we treat $h(t)$, which is assumed to be non-negative for all $t \geq 0$, as the probability density function (p.d.f.), i.e., $\int_0^\infty h(t) = 1$, the Elmore delay, T_d, is essentially the mean under the p.d.f. of $h(t)$, the impulse response of the network

$$T_d = m_1 = \int_0^\infty h(t)t\mathrm{d}t, \tag{2.16}$$

which actually is the first-order moment m_1.

It can be shown that for an ideal delay network or a network whose impulse response is symmetric, the Elmore delay is exactly the actual 50% delay of the network. For practical networks, whose responses are always skewed as shown in Figure 2.2, the Elmore delays are just an estimation. Actually Gupta and Pileggi *et al.* proved that the Elmore delay is the upper bound for general RC circuits [48].

The Elmore delay was first introduced by Elmore in 1948 for estimating the delay of active circuits. It was popularized by Penfield and Rubinstein [100] as it can be computed directly and efficiently for RC trees by using the R and C values

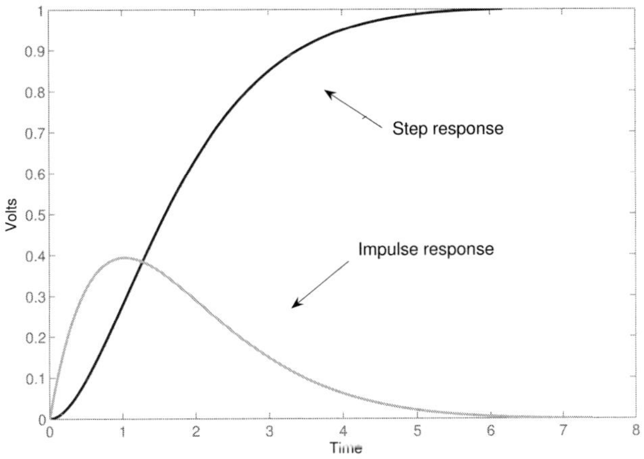

Figure 2.2 The unit impulse and unit step responses.

from the tree circuits in a linear time. Because of its high fidelity as a delay metric, it was intensively used for the delay and timing estimation of interconnects during the physical layout design processes [23].

2.2 Moment computation in MNA formulation

In this section, we present the method for efficient computation of moments using classic circuit analysis techniques. We start with the modified nodal analysis (MNA) formulation of the general RLC linear circuits and derive the recursive moment computation formula, which is the most critical step for all the moment-matching methods and Krylov subspace projection-based model order reduction methods.

2.2.1 *Recursive moment computation*

For a general linear network, we can apply modified nodal analysis to formulate it in the state space equation form

$$\begin{aligned} G\mathbf{x}(t) + C\dot{\mathbf{x}}(t) &= Bu(t) \\ \mathbf{y}(t) &= L^T\mathbf{x}(t), \end{aligned} \tag{2.17}$$

where G and C are the conductive and storage element matrices; B and L are the input and output positions matrices; and state variables x can be nodal voltage or branch currents of the linear circuit.

Upon applying the Laplace transformation of the state equation, we have the state equations in the s domain

$$\begin{aligned} GX(s) + sCX(s) - CX(0) = BU(s) \\ Y(s) = L^T X(s). \end{aligned} \tag{2.18}$$

Assuming the initial condition is zero, $X(0) = 0$, and the impulse response ($U(s) = 1$) is applied, the state equation will become

$$\begin{aligned} (G + sC)X(s) = B \\ Y(s) = L^T X(s) \end{aligned} \tag{2.19}$$

Expanding $X(s)$ using Taylor's series at $s = 0$, we obtain

$$(G + sC)(\mathbf{x}_0 + \mathbf{x}_1 s + \mathbf{x}_2 s^2 + \cdots + \mathbf{x}_q s^q + \cdots) = B. \tag{2.20}$$

We then obtain the state moment computation formula in a recursive form

$$\begin{aligned} &\mathbf{x}_0 = G^{-1}B \\ &\mathbf{x}_1 = -G^{-1}C\mathbf{x}_0 \\ &\ldots \\ &\mathbf{x}_i = -G^{-1}C\mathbf{x}_{i-1} \text{ for } i > 0, \end{aligned} \tag{2.21}$$

Notice that G^{-1} here means we solve $G\mathbf{x} = \mathbf{b}$ and G^{-1} is used for solving for all the moments. Numerically, we only need to perform one LU decomposition of $G = LU$ and then use the L and U matrices to solve for all moments sequentially. Typically, we only need a few orders, say q, of moments for achieving the required accuracy in terms of transient waveforms for RC or RLC circuits, where $q \ll n$ and n is the size of state vector $\mathbf{x}$. As a result, transient analysis using the moment method is much faster than integration-based numerical analysis as it is independent of time steps and time intervals.

For the output moments, which actually are the moments of the transfer function:

$$H(s) = L^T(G + sC)^{-1}B. \tag{2.22}$$

The moments m_i are related to the state moment by

$$\mathbf{m}_i = L^T \mathbf{x}_i. \tag{2.23}$$

As a result, the transfer function moment $\mathbf{m}_i$ can be directly computed in a recursive way

$$\begin{aligned} &\mathbf{x}_0 = G^{-1}B;\ \mathbf{m}_0 = L^T\mathbf{x}_0 \\ &\mathbf{x}_1 = -G^{-1}C\mathbf{x}_0;\ \mathbf{m}_1 = L^T\mathbf{x}_1 \\ &\ldots \\ &\mathbf{x}_i = -G^{-1}C\mathbf{x}_{i-1};\ \mathbf{m}_i = L^T\mathbf{x}_i \text{ for } i > 0, \end{aligned} \tag{2.24}$$

In general, the ith block moment is given by

$$\mathbf{m}_i = L^T(-G^{-1}C)^i G^{-1}B. \tag{2.25}$$

2.3 Asymptotic waveform evaluation

Asymptotic waveform evaluation (AWE) is an efficient frequency-domain analysis approach and was proposed in 1990 by L. Pileggi [91, 92]. It basically combines the fast recursive moment computation methods presented in Subsection 2.2.1 with *Pade* approximation to compute the poles and residues of the order truncated transfer functions. We first present the Pade approximation method.

2.3.1 *Pade approximation*

The idea of Pade approximation is to approximate a transfer function $\mathbf{H}(s)$ by an order-limited rational function $\mathbf{H}_q(s)$, where q is the order.

Specifically, after the moments are generated, a general multi-input multi-output (MIMO) transfer function $\mathbf{H}(s)$ is represented as a Taylor series expansion form or the block moment form

$$\mathbf{H}(s) = \mathbf{m}_0 + \mathbf{m}_1 s + \mathbf{m}_2 s^2 + \cdots, \tag{2.26}$$

where $\mathbf{m}_i$ is the ith block moment of the circuit, and

$$\mathbf{m}_i = L^T \mathbf{x}_i, \tag{2.27}$$

where $\mathbf{x}_i$ is the ith order state block moment vector. Once the moment expansion is available, a Pade approximation is calculated. For a qth order approximation, $2q$ moments must be computed.

Without losing generality, we consider a single-input single-output transfer function. Consider the transfer function at entry (p, q) and let $m_i = \mathbf{m}_{i,pq}$ for $i = 0, 1, 2, \cdots$, the scalar moment expansion then can be written as

$$Y(s) = m_0 + m_1 s + m_2 s^2 + \cdots + m_{2q-1} s^{2q-1}. \tag{2.28}$$

Then we can use a qth Pade approximation rational function $H_q(s)$ to match $H(s)$,

$$H(s) = \frac{P(s)}{Q(s)} = \frac{a_0 + a_1 s + a_2 s^2 + \cdots + a_{q-1} s^{q-1}}{1 + b_1 s + b_2 s^2 + \cdots + b_q s^q}, \tag{2.29}$$

such that they agree on the first $2q$ terms in the moment form, i.e.,

$$H(s) = H_q(s) + O(s^{2q}). \tag{2.30}$$

To compute the coefficients a_i and b_i from (2.29), we have

$$P(s) = H_q(s) Q(s), \tag{2.31}$$

or

$$\sum_{k=0}^{q-1} a_k s^k = (\sum_{k=0}^{2q-1} m_k s^k)(\sum_{k=0}^{q} b_k s^k). \tag{2.32}$$

Here $b_0 = 1$. The equation can be solved for different powers of s separately. If we compare the coefficients of s^q on the left-hand side and the right-hand side of (2.32), we have

$$0 = m_0 b_q + m_1 b_{q-1} + \cdots + m_{q-1} b_1 + m_q. \tag{2.33}$$

Doing this for the coefficients of s^q to s^{2q-1} yields a set of linear equations

$$\begin{bmatrix} m_0 & m_1 & \cdots & m_{q-1} \\ m_1 & m_2 & \cdots & m_q \\ \vdots & \vdots & \ddots & \vdots \\ m_{q-1} & m_q & \cdots & m_{2q-2} \end{bmatrix} \begin{bmatrix} b_q \\ b_{q-1} \\ \vdots \\ b_1 \end{bmatrix} = - \begin{bmatrix} m_q \\ m_{q+1} \\ \vdots \\ m_{2q-1} \end{bmatrix}. \tag{2.34}$$

Through (2.34), we can solve for all the coefficients b_i of $Q(s)$.

To solve for the coefficients a_k of numerator $Q(s)$, we solve for the following equations, which are obtained by comparing the coefficients of s^q on the left-hand side and the right-hand side of (2.32) from s^0 to s^{q-1}, which yields

$$\begin{bmatrix} a_0 \\ a_1 \\ \vdots \\ a_{q-1} \end{bmatrix} = \begin{bmatrix} m_0 & 0 & 0 & \cdots & 0 \\ m_1 & m_0 & 0 & \cdots & 0 \\ \cdots & \cdots & \cdots & \cdots & \cdots \\ m_{q-1} & m_{q-2} & m_{q-3} & \cdots & m_0 \end{bmatrix} \begin{bmatrix} 1 \\ b_1 \\ \vdots \\ b_{q-1} \end{bmatrix}. \tag{2.35}$$

2.3.2 *Partial fraction decomposition and time-domain response*

After the Pade approximation, we obtain the order-reduced order transfer function in (2.29). To compute the transient response waveforms, we need to derive the partial fraction form from the rational function form.

Partial fraction decomposition of a transfer function $H(s)$ is to represent it in the following form,

$$\begin{aligned} H(s) &= \frac{k_0}{s - p_0} + \frac{k_1}{s - p_1} + \cdots + \frac{k_p}{s - p_q}. \\ &= \sum_{i=0}^{q} \frac{k_i}{s - p_i}, \end{aligned} \tag{2.36}$$

where p_i and k_i are the poles and residues of the transfer function.

From this representation of $H(s)$, the time-domain impulse response $h(t)$ can be computed by the closed form expression

$$h(t) = \sum_{i=0}^{q} k_i e^{p_i t} u(t), \tag{2.37}$$

where $u(t)$ is the unit step function. Once the impulse response poles and residues are known, the responses for some ideal signals such as the step and limited ramp inputs can be expressed easily.

For instance, for unit step input $U(s) = 1/s$, the output response in the s-domain is

$$\begin{aligned} Y(s) &= H(s)U(s) \\ &= H(s)\frac{1}{s} \\ &= \sum_{i=0}^{q} \frac{k_i}{(s-p_i)s} \\ &= \sum_{i=0}^{q} \frac{k_i}{p_i}\left(\frac{1}{s-p_i} - \frac{1}{s}\right), \end{aligned} \tag{2.38}$$

and the corresponding time-domain response, $y(t)$, is

$$y(t) = \left[\sum_{i=0}^{q} \frac{k_i}{p_i}(\mathrm{e}^{p_i t} - 1)\right] u(t). \tag{2.39}$$

2.3.3 *Derivation of poles and residues*

To derive the partial fraction representation of a rational transfer function $H(s)$, we need to compute the poles p_i and residues k_i.

The poles are actually the roots of the denominator polynomials of $Q(s)$ in (2.29)

$$Q(s) = 1 + b_1 s + b_2 s^2 + \cdots + b_q s^q,$$

and can be obtained by many numerical root computation processes [44].

To compute the residues k_i, there are two methods. The first method requires that we know both numerator $P(s)$ and denominator $Q(s)$ (i.e., their coefficients). While the second method does not need to compute the numerator polynomial $P(s)$, it can compute the residues directly once the poles are available.

We first present the first method. Notice that for the residues k_j in (2.36), we have

$$H(s) = \frac{P(s)}{\prod_{i=0}^{q}(s-p_i)} = \sum_{i=0}^{q} \frac{k_i}{s-p_i}.$$

If we multiply factor $(s-p_j)$ on both sides, we obtain

$$\frac{P(s)}{\prod_{i=0,i\neq j}^{q}(s-p_i)} = \sum_{i=0,i\neq j}^{q} \frac{k_i}{s-p_i}(s-p_j) + k_j.$$

To derive k_j, we substitute s for p_j in the above equation, thus

$$k_j = \frac{P(p_j)}{\prod_{i=0,i\neq j}^{q}(p_j-p_i)}. \tag{2.40}$$

In the second method for computing residues k_i, we first expand each of the partial fraction terms in (2.36):

$$H(s) = \sum_{i=0}^{q} \frac{k_i}{s - p_i} = \sum_{i=0}^{q} \left[-\frac{k_i}{p_i} \left(1 + \frac{s}{p_i} + \frac{s^2}{p_i^2} + ... \right) \right]. \tag{2.41}$$

Comparing coefficients of (2.41) with that of $H(s)$ in its moment expansion form

$$H(s) = m_0 + m_1 s + m_2 s^2 + \cdots + m_{2q-1} s^{2q-1},$$

we can arrive at

$$\begin{aligned} &-(\tfrac{k_1}{p_1} + \tfrac{k_2}{p_2} + \cdots + \tfrac{k_q}{p_q}) = m_1, \\ &-(\tfrac{k_1}{p_1^2} + \tfrac{k_2}{p_2^2} + \cdots + \tfrac{k_q}{p_q^2}) = m_2, \\ &\vdots \\ &-(\tfrac{k_1}{p_1^{2q}} + \tfrac{k_2}{p_2^{2q}} + \cdots + \tfrac{k_q}{p_q^{2q}}) = m_{2q}. \end{aligned} \tag{2.42}$$

Actually, since the number of unknown k_i is q, we only need to select q equations from (2.42) to solve for $k_i, i = 1, ..., q$.

We summarize the steps involved in AWE using moment matching in the following algorithm:

Asymptotic waveform evaluation

1. Compute $2q$ moments using (2.21), choice of q depends on accuracy requirement. In practice, $q < 5$ is typically used.
2. Obtain the coefficients of denominator polynomials $Q(s)$ and coefficients of numerator polynomials $P(s)$ (if using the first method to compute residues) in (2.29) by solving (2.34) and (2.35).
3. Obtain the roots (thus the poles $p_i, i = 1, ..., q$) of the denominator polynomials $Q(s)$.
4. Find the corresponding residues $k_q, i = 1, ..., q$ using (2.40) or (2.42).
5. Compute the time domain responses for given inputs using closed form expressions for ideal input signals or recursive convolution for general input signals.

One important observation from (2.42) is that the moment m_i is the power function of poles $m_i(p_1, p_2, ...p_q)$. As a result, the higher order (less dominant) poles, which have large magnitudes, will soon be lost numerically in the moments and high order moments will approximate the dominant poles (small poles). Practically, this reflects the fact that explicit moment methods like AWE cannot reliably obtain high-order poles and AWE can easily generate unstable positive poles when high-order moments are computed [91]. Typically, the reliable orders that can be obtained by AWE are approximately 5-6 based on our experiments. That implies two or three poles can be reliably computed by AWE.

To mitigate this problem, many approaches have been proposed in the past. Examples are frequency scaling, frequency shifting [92] and multiple point expansions such as complex frequency hopping (CFH) [19]. In this book, we present two recent developments for improving the numerical stability of direct moment matching in the sequel.

2.3.4 Multi-node moment matching (MMM)

As mentioned in the previous section, the major issue with direct moment-matching methods like AWE is that the higher order moments will become less accurate: higher order moments are power functions of poles as shown in (2.42). Higher order pole information will soon be lost numerically. One effective way to partially mitigate this problem is by means of the multi-node moment matching technique (MMM), which can obtain more numerical stable estimation of the system poles [54].

The main idea of the MMM method is to estimate the poles by using the moment information from different nodes or from different input stimulus instead of using a single node with a single stimuli as done in the traditional moment-matching method [91]. The rational behind this method is that poles from different nodes and from different input stimulus are the same.

By using MMM, we only need fewer orders of moments to compute the same number of poles. For instance, if moments from p different nodes with a single input are used only, we can get the p poles by using just $p+1$ moments. If k inputs are allowed, then we can get the p poles by just using $p/k+1$ moments for each input with about the same computing cost. Since the pole information is better preserved numerically in lower-order moments, as shown in (2.42), we can obtain a much better and numerically stable estimation of the poles.

Specifically, we start with a linear dynamic system with n states:

$$\dot{\mathbf{x}} = \mathbf{A}\mathbf{x} + \mathbf{B}u, \tag{2.43}$$

where $\mathbf{A}$ is the $n \times n$ original system and $\mathbf{B}$ is the $n \times 1$ position matrix.

Assume that there exists a reduced system, $\mathbf{A}_{q\times q}$, of order q to be determined that approximates the original circuit by simultaneously matching the moments of the selected q state variables $\mathbf{x}$. This system is given by

$$\dot{\mathbf{x}} = \mathbf{A}_{q\times q}\mathbf{x} + \mathbf{B_q}u, \tag{2.44}$$

where $\mathbf{A}_{q\times q}$ is a $q \times q$ reduced system and $\mathbf{B_q}$ is a $q \times 1$ reduced position matrix.

We select q nodes, which have one-one correspondence with the q variables selected in $\mathbf{A}_{q\times q}$, and their corresponding moments in the original circuit become the moment vectors $\mathbf{m_0}$, $\mathbf{m_1}$,$\mathbf{m_2}$,...,$\mathbf{m}_{q+1}$. According to (2.43), we have

$$\begin{aligned} s[\mathbf{m_0} + \mathbf{m_1}s + \mathbf{m_2}s^2 + ... + \mathbf{m_q}s^q + ...] = \\ \mathbf{A}_{q\times q}[\mathbf{m_0} + \mathbf{m_1}s + \mathbf{m_2}s^2 + ... + \mathbf{m_q}s^q + ...] + \mathbf{B_q}u. \end{aligned} \tag{2.45}$$

By comparing the coefficients of equal powers of s in both sides, we have the following equations:

$$\begin{aligned}
\mathbf{B_q} &= -\mathbf{A}_{q\times q}\mathbf{m}_0, \\
\mathbf{m}_0 &= \mathbf{A}_{q\times q}\mathbf{m}_1, \\
\mathbf{m}_1 &= \mathbf{A}_{q\times q}\mathbf{m}_2, \\
\ldots \\
\mathbf{m}_{q-1} &= \mathbf{A}_{q\times q}\mathbf{m}_q.
\end{aligned} \tag{2.46}$$

Excluding the first equation, the equations above can be put into a matrix form as

$$\mathbf{A}_{q\times q}[\mathbf{m}_1, \mathbf{m}_2, ..., \mathbf{m}_q] = [\mathbf{m}_0, \mathbf{m}_1, \mathbf{m}_2, ..., \mathbf{m}_{q-1}]. \tag{2.47}$$

As a result, we can compute $\mathbf{A}_{q\times q}$ as

$$\mathbf{A}_{q\times q} = [\mathbf{m}_0, \mathbf{m}_1, \mathbf{m}_2, ..., \mathbf{m}_{q-1}][\mathbf{m}_1, \mathbf{m}_2, ..., \mathbf{m}_q]^{-1}. \tag{2.48}$$

Then the eigenvalues of $\mathbf{A}_{q\times q}$, which can be obtained by performing the eigendecomposition on $\mathbf{A}_{q\times q}$, are the reciprocals of q dominant poles of the original system in (2.43). Notice that only $q + 1$ moments are required to compute the q dominant poles when we select q nodes.

If we use the moments from different inputs, then the order of moments used in (2.47) can be further reduced, as shown in [54].

2.3.5 *Projection-based methods for pole computation*

Another way of finding poles in a numerically reliable way is by means of projection-based model order reduction methods, where poles are computed from the eigendecomposition of the reduced system.

This method makes sense if we are interested in the fast transient response by using explicit moment-matching methods instead of model order reduction of the systems. Explicit circuit moment matching can be transformed to the time domain waveforms very easily without simulating the reduced models. Hence, explicit moment matching is still appealing if poles can be computed reliably.

Krylov subspace projection-based methods will be discussed in detail in the following sections. Here, we just briefly mention how the methods can be used to generate the dominant poles of a system.

In the Krylov-subspace projection-based model order reduction process, moment vectors generated recursively in (2.21) are first orthonormalized during the generation process and then used to build a projection matrix. The projection matrix is then used to reduce the original circuit matrices by congruence transformation, which ensures that the reduced system is passive (thus stable) if original matrix are formulated properly [63, 85]. By using this method, we only require q moments to find q poles. Such a method guarantees that all the poles computed are stable (less than zero in their real part) [63, 85] owing to the nature of congruence transformation and the MNA formulation of the original RLC circuit matrices.

Specifically, we obtain the first q moment vectors through (2.21). Then we form the following $n \times q$ matrix where each moment vector is a column.

$$M = [\mathbf{m}_0, \mathbf{m}_1, \cdots, \mathbf{m}_{q-1}]_{n\times q}. \tag{2.49}$$

where $q \ll n$, and n is the number of state variables (nodes) in the original circuit and also the dimension of the moment vectors. Then we orthonormalize M into an $n \times q$ projection matrix V, such that the columns in V are mutually orthogonal, i.e., $v_i^T v_j = \delta_{ij}, i \neq j$. Such an orthogonalization process can be simply carried out by using the standard *Gram–Schmidt* method [44]. Actually, the moments can be orthonormalized during the generation process, as in methods such as the Arnoldi or Lanczos methods, which are numerically more stable than the standard Gram-Schmidt method [44]. We will discuss the subspace projection methods in detail in the following sections.

Once we have obtained the projection matrix V, the original circuit matrix $\mathbf{G}$ and $\mathbf{C}$ in (2.17) can be reduced to two $q \times q$ order reduced matrices by the *congruence transformation*:

$$\hat{\mathbf{G}} = V^T\mathbf{G}V, \quad \hat{\mathbf{C}} = V^T\mathbf{C}V. \tag{2.50}$$

After this reduction process, the eigenvalues of matrix $\hat{\mathbf{G}}^{-1}\hat{\mathbf{C}}$ will be related to the dominant poles we are looking for as:

$$p_i = -\frac{1}{\lambda_i}, \tag{2.51}$$

where p_i and λ_i are the ith pole and eigenvalue. This can be easily obtained by performing the eigendecomposition of $\hat{\mathbf{G}}^{-1}\hat{\mathbf{C}}$. Once all the poles are computed, we then compute the residues at any node using aforementioned residue computation methods.

We want to stress that the major difference of the projection methods used for pole computations here and the full-blown subspace projection-based model order reduction methods like PRIMA [85] is that we only need to compute the dominant poles here instead of the reduced system. Since the dominant poles are shared by all the transfer functions, we only need one transfer function to compute them. As a result, we can perform the projection-based model order reduction assuming one terminal. In this case, we need to make sure that the order of moments will be the same or larger than the required number of poles. If we use multiple terminals, then we end up with block moments. But we only need a few orders of block moments so that the reduced models have larger number of poles than is required. This case is similar to the multiple-input matching in the aforementioned MMM method.

The bottom line is that the pole-computation reduction process (its CPU cost) does not depend on the number of terminals of the systems, which is not the case for full-blown Krylov subspace projection-based model order reduction methods to be discussed later.

We notice that the Krylov subspace projection-based methods for pole computation have been used to compute the transient thermal profile using explicit moment matching in the microprocessor architecture level [74].

2.4 Projection-based model order reduction methods

In this section, we present projection-based model order reduction techniques, which are widely used for parasitic interconnect circuit macromodeling and reductions. We mainly focus on model order reduction approaches based on Krylov subspace projection methods.

2.4.1 Framework of projection-based model order reduction

We consider a general linear time-invariant state-space model with only one input and one output (we will extend our discussion to multi-input, multi-output cases later),

$$\begin{aligned} \dot{\mathbf{x}}(t) &= A\mathbf{x}(t) + \mathbf{b}u(t) \\ y(t) &= \mathbf{l}^T\mathbf{x}(t), \end{aligned} \tag{2.52}$$

where u is the input variable, y is the output variable, A is an $n \times n$ matrix, and $\mathbf{b}$ is an $n \times 1$ matrix. $\mathbf{x}$ is an $n \times 1$ vector of state variables. Then the transfer function from $u(t)$ to $y(t)$ can be given as

$$H(s) = \mathbf{l}^T(sI - A)^{-1}\mathbf{b}. \tag{2.53}$$

Typically, the number of state variables, n, is very large so that the simulation and synthesis of the whole systems are very slow. We want to build a much smaller system, such that the transient response $y(t)$ to some given input signal $u(t)$ is approximate to that by the original system. One question we may ask is which parts of the system can be discarded without changing the transfer function $H(s)$ significantly. The concepts of controllability and observability can give good answers: uncontrollable and unobservable parts of the system can be removed without affecting the transfer function [15].

To this end, we can perform the state transformation $\mathbf{x} = T\mathbf{z}$, where T is the matrix of eigenvectors of A and we assume that the eigenvalues of A are denoted $\lambda_1, ..., \lambda_n$ and they are simple and unique. Then (2.52) becomes

$$\begin{aligned} \dot{\mathbf{z}}(t) &= T^{-1}AT\mathbf{z}(t) + T^{-1}\mathbf{b}u(t) = \begin{bmatrix} \lambda_1 & & \\ & \ddots & \\ & & \lambda_n \end{bmatrix} \mathbf{z}(t) + \begin{bmatrix} \bar{b}_1 \\ \vdots \\ \bar{b}_n \end{bmatrix} u(t), \\ y(t) &= \mathbf{l}^T T\mathbf{z}(t), \end{aligned} \tag{2.54}$$

Figure 2.3 illustrates the eigen-decomposed system, which consists of n independent transfer paths. If $\bar{b}_i$ is zero, then the state variable z_i is not controllable and can be removed; Let $\bar{\mathbf{c}} = \mathbf{l}^T$, if $\bar{c}_i$ is zero, then the state variable z_i is unobservable and can be removed. So the key issue of model order reduction is to remove uncontrollable or unobservable parts or practically weakly controllable or observable parts. One way to perform the direct state removal is by using a balancing and truncation method based on balanced realization instead of the diagonal representation [81], which will be discussed in detail in Chapter 3.

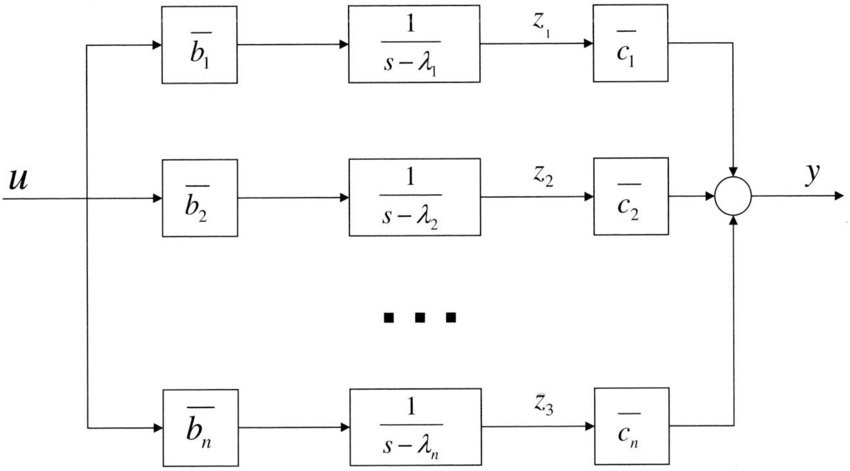

Figure 2.3 Block diagram of (2.54).

Another way to do the order reduction is by means of projection. Specifically, if we assume that the first q state variables in (2.54) to be preserved and the remaining variables are to be removed, we can form a new *projection* matrix T_p such that T_q consists of the first q left columns of T; we have

$$\begin{aligned} \dot{\mathbf{z}}(t) &= T_q^{-1}AT_q\mathbf{z}(t) + T_q^{-1}\mathbf{b}u(t) \\ y(t) &= \mathbf{l}^T T_q\mathbf{z}(t), \end{aligned} \tag{2.55}$$

Notice that $A_r = T_q^{-1}AT_q$ is a $q \times q$ matrix, and $\mathbf{b}_r = T_q^{-1}\mathbf{b}$ is a $q \times 1$ matrix. Thereby, the original state space of dimension n is projected to the space of the reduced model with dimension q.

In general, for a single-input and single-output system (SISO):

$$\begin{aligned} E\dot{\mathbf{x}}(t) &= A\mathbf{x}(t) + \mathbf{b}u(t) \\ y(t) &= \mathbf{l}^T\mathbf{x}(t), \end{aligned} \tag{2.56}$$

instead of using biorthogonal matrices T_q and T_q^{-1}, we can use other $n \times q$ matrices V and W for the projection. As a result, we may have more general projection reduction methods in terms of V and W for a single-input and single-output system,

$$\begin{aligned} W^T EV\dot{\mathbf{z}}(t) &= W^T AV\mathbf{z}(t) + W^T\mathbf{b}u(t) \\ y(t) &= \mathbf{l}^T V\mathbf{z}(t). \end{aligned} \tag{2.57}$$

In projection theory, the $q \times q$ matrix $W^T AV$ can be interpreted as projecting A onto the subspace $\mathcal{L}$ spanned by V, and orthogonal to the subspace $\mathcal{M}$ spanned by W [102].

The projection operation approximates a solution to a system from a search subspace $\mathcal{L}$ of dimension q, so that q constraints are met at the same time. The projection operations are always involved with two spaces. If $\mathcal{L} = \mathcal{T}$, the projection is said to be *orthogonal*. Otherwise, the projection is *oblique*.

For a linear dynamic system, the resulting projection operation will lead to matrix transformation as shown in (2.57). To see this in terms of a general projection operation, we first project the state variable x into the subspace $\mathcal{L}$ represented by V, i.e., $\text{span}(V) = \mathcal{L}$, we have

$$\mathbf{x} = V\mathbf{z}. \tag{2.58}$$

Since (2.58) is an approximate solution in subspace $\mathcal{L}$, we require that the residues in (2.52) will be orthogonal to the subspace $\mathcal{M} = \text{span}(W)$, which satisfies the so-called *Petrov–Galekin* conditions. The residues for (2.52) can be written as (assuming the impulse response $u(s) = 1$)

$$\mathbf{e}_r = \mathbf{b} + (A - sE)\mathbf{x}. \tag{2.59}$$

If we multiply (2.59) by W^T and replace $\mathbf{x}$ with $V\mathbf{z}$, we have

$$W^T\mathbf{e}_r = W^T\mathbf{b} + W^T(A - sE)V\mathbf{z} = 0, \tag{2.60}$$

which is the same state transfer function we have in (2.57).

One question left is how to select the two subspaces and their corresponding projection matrices V and W. Krylov subspaces are the most useful and popular ones owing to their moment matching connection with the original system [32].

2.4.2 *Krylov subspaces*

A subset of a vector space is called a vector subspace or subspace. The subspace is uniquely defined by a set of vectors $V = \{v_1, v_2, ..., v_n\}$. The set of all linear combinations of these vectors is referred to as the span of V.

$$\text{span}\{V\} = \text{span}\{v_1, v_2, ..., v_n\}, \tag{2.61}$$

$$= \left\{ x | x = \sum_{i=1}^{n} \alpha_i v_i \right\}, \tag{2.62}$$

where α_i are real numbers. If v_i are linearly independent, then each vector of $\text{span}\{V\}$ admits a unique expression as a linear combination of v_i. The set V is then called a basis of the subspace $\text{span}\{V\}$.

Concept of Krylov subspaces

For $n \times n$ matrix A and a vector $\mathbf{b}$, the Krylov subspace $K_q(A, \mathbf{b})$ is defined as

$$K_q(A, \mathbf{b}) = \text{span}\{\mathbf{b}, A\mathbf{b}, A^2\mathbf{b}, ..., A^q\mathbf{b}\}, \tag{2.63}$$

where q is a given positive integer.

For an $n \times q$ matrix T_q, whose columns form bases for the subspace spanned by the $K_q(A, \mathbf{b})$,

$$\text{colsp}T_q = K_q(A, \mathbf{b}), \tag{2.64}$$

where colsp(G) indicates the column space of G. With this definition, the two projection matrices V and W determining the reduced order model for (2.56) are chosen as follows:

$$\begin{aligned} \text{colsp}V &= K_q(A^{-1}E, A^{-1}\mathbf{b}) &= \text{span}\{A^{-1}\mathbf{b}, ..., (A^{-1}E)^{q-1}A^{-1}\mathbf{b}\}, \\ \text{colsp}W &= K_q(A^{-T}E^T, A^{-T}\mathbf{l}) &= \text{span}\{A^{-T}\mathbf{l}, ..., (A^{-T}E^T)^{q-1}A^{-T}\mathbf{l}\}, \end{aligned} \tag{2.65}$$

where $A^{-T} = (A^{-1})^T$ and V and W are both rank q. In this book, we call V the *output* Krylov subspace and W the *input* Krylov subspace.

With this choice, the reduction method is called a *two-side* method as both input and output relations of the systems are involved. If only one of the two matrices V or W is chosen and the other one is chosen arbitrarily such that W^TAV is non-singular, then the method is called a *one-side* MOR method.

If we compute the V and W directly based on their definitions in (2.65), we will find that the numerical calculation of the matrix-vector products involved in the subspaces turns out to be unstable. This is no surprise as $\{(A^{-1}E)^{q-1}A^{-1}\mathbf{b}\}$ are the circuit moments, which will lose the small eigen value information numerically, as the order q becomes large, as shown in the previous section. The reduction process may lead to unstable poles as a result. Practically, the Krylov spaces can be obtained by numerical stable processes like Arnoldi and Lanczos algorithms [44], which are to be discussed in the next subsections.

Moment connection of Krylov subspaces

One important property of the Krylov subspace is that the output responses of reduced models match that of the original circuits in terms of moments [32]. To see this, we rewrite the transfer function for the state space equation in (2.56) as

$$\begin{aligned} H(s) &= \mathbf{l}^T(sE - A)^{-1}\mathbf{b} \\ &= \mathbf{l}^T(s\bar{A} - I)^{-1}\bar{\mathbf{b}}, \end{aligned} \tag{2.66}$$

where $\bar{A} = A^{-1}E$ and $\bar{\mathbf{b}} = A^{-1}\mathbf{b}$. As a result, we can obtain the coefficients of the Taylor series of (2.66) around $s_0 = 0$,

$$\begin{aligned} H(s) &= \mathbf{l}^T(s\bar{A} - I)^{-1}\bar{\mathbf{b}}, \\ &= -\mathbf{l}^T\bar{\mathbf{b}} - \mathbf{l}^T\bar{A}\bar{\mathbf{b}}s, \cdots, -\mathbf{l}^T\bar{A}^i\bar{\mathbf{b}}s^i \cdots \\ &= -\mathbf{l}^TA^{-1}\mathbf{b} - \mathbf{l}^T(A^{-1}E)A^{-1}\mathbf{b}s, \cdots, -\mathbf{l}^T(A^{-1}E)^iA^{-1}\mathbf{b}s^i \cdots . \end{aligned} \tag{2.67}$$

In the two side methods, the first two $2q$ moments of the original and the reduced model match. In one-side methods, only q moments of the original and the reduced model match. In the sequel, we first present the moment matching theorem and then we give a simple proof for $s_0 = 0$ case. A more general proof can be found in [45].

Theorem 2.1 (Moment matching connection of one-side Krylov subspace). *For the two Krylov subspaces* colsp$V = K_q(A^{-1}E, A^{-1}\mathbf{b})$ *and* colsp$W = K_q(A^{-T}E^T, A^{-T}\mathbf{l})$ *defined in* (2.65), *For the one-side Krylov subspace method, the reduced system obtained by projection reduction matches q moments of the transfer function of the original system defined in* (2.66).

The proof can be found at the Section 2.8. If we use both Krylov spaces colsp$V = K_q(A^{-1}E, A^{-1}\mathbf{b})$ and colsp$W = K_q(A^{-T}E^T, A^{-T}\mathbf{l})$, then we have the following result:

Corollary 2.1 (Moment matching connection of two-side Krylov aubspace methods). *If both subspaces V and W are used, the reduced model matches $2q$ moments of the transfer function of the original system defined in* (2.66).

See Section 2.8 for a simple proof. In the one-side Krylov methods, since we only use one Krylov subspace, say, V, a good choice for W is $V = W$. As a result, the matrix transformation W^TAV become V^TAV, which is called *congruence* transformation. It was shown that congruence transformation can preserve the passive property of the original systems if the state space matrix is formed in a passive form. We will discuss this in the later subsection. Here we want to present some result regarding the symmetric systems.

Corollary 2.2 (Moment matching connection for symmetric systems). *For a symmetric system* $\{A, E, \mathbf{b}, \mathbf{l}\}$ *in* (2.56) *with the same input and output ports , i.e., A and E are symmetric and* $\mathbf{b} = \mathbf{l}$, *for one-side Krylov subspace methods with congruence transformation (i.e., $V = W$), the reduced model matches $2q$ moments of the transfer function of the original system defined in* (2.66).

Proof: This corollary can easily be proved by noticing that, in this case, the two subspaces colsp$V = K_q(A^{-1}E, A^{-1}\mathbf{b})$ and colsp$W = K_q(A^{-T}E^T, A^{-T}\mathbf{l})$ are the same because of the symmetric nature of the system. Also, since we make $V = W$, such reduction is equivalent to the two-side method and, thus, $2q$ moments are matched, based on Corollary 2.1. QED.

The significance of Corollary 2.1 is that for many practical RC circuits, whose MNA matrices are symmetric, using the one-side method such as the Arnoldi method can lead to more accurate results and the reduced models are also passive. For general RLC circuits, their matrices are not symmetric, but by splitting the projection matrices, which makes the matrices subblock-level symmetric, the reduction process can still lead to double accuracy (matching $2q$ instead of q moments) as shown in [37] and in Chapter 8 for general structure-preserved model order reduction methods. Another way is by means of the second order formulation of RLC circuits, which is also symmetric and can lead to $2q$ moment matching.

Note that instead of $s_0 = 0$, the expansion can be carried out at another point $s_0 \neq 0$. Then, with slight modification of the Krylov subspaces, the moments can still be matched here.

In the following subsections, we show two numerically stable algorithms, the *Arnoldi* method and the *Lanczos* method to find the Krylov subspaces. The Arnoldi method is an one-side method while Lanczos method is a two-side method.

The fundamental idea is to orthonormalize the basis vectors in the Krylov subspaces defined in colsp$V = K_q(A^{-1}E, A^{-1}\mathbf{b})$ or colsp$W = K_q(A^{-T}E^T, A^{-T}\mathbf{l})$. Such orthogonal vectors contain much less numerical noise compared with the circuit moments as lower-order moment vectors are subtracted during the orthonormalization process.

2.4.3 *Arnoldi algorithms*

To avoid numerical problems when constructing the Krylov subspace, it is advantageous to build the orthogonal basis for a given subspace. The Arnoldi method is the classic method to find a set of orthonormal vectors as a basis for a given Krylov subspace as proposed by W.E. Arnoldi in 1951 [8].

Specifically, given Krylov subspace $K_q(A, \mathbf{b})$, the Arnoldi method using the *modified Gram–Schmidt* orthogonalization [36, 102, 104] is as follows:

ARNOLDI ALGORITHM $(A, \mathbf{b})$
Compute $\mathbf{v}_1 = \frac{\mathbf{b}}{\|\mathbf{b}\|_2}$ —- # normalize **b**
For $j = 1, ..., q$ Do
Compute $\mathbf{w}_j = A\mathbf{v}_j$ —- # calculate the next vector
For $i = 1, ..., j$ Do
$h_{ij} = \mathbf{v}_i^T \mathbf{w}_j$
$\mathbf{w}_j = \mathbf{w}_j - h_{ij}\mathbf{v}_i$ —- # orthogonalization
EndDo
If $\mathbf{w}_i = 0$, stop, otherwise $h_{j+1,j} = \|\mathbf{w}_j\|_2$
$\mathbf{v}_{j+1} = \frac{\mathbf{w}_j}{h_{j+1,j}}$ —- # normalization
EndDo

Figure 2.4 Arnoldi method based on modified Gram–Schmidt orthonormalization for SISO systems.

After the Arnoldi process, we obtain a new set of vectors $V = \{\mathbf{v}_1, \mathbf{v}_2, ..., \mathbf{v}_q\}$ and new orthogonal projection matrix V

$$V^T V = I, \tag{2.68}$$

whose columns are the basis for the given Krylov subspace. Furthermore,

$$V^T A V = H, \tag{2.69}$$

where H is a $q \times q$ matrix, called a *Hessenberg* matrix , in which $h_{ij} = 0$ for $i > j + 1$ for all $1 \le i,\ j \le q$.

The Arnoldi method presented in Figure 2.4 is just for a single-input and single-output system (SISO). For the multi-input and multi-output case, block Arnoldi methods can be applied [8, 14].

2.4.4 Lanczos algorithms

The Lanczos algorithm is a well known two-side Krylov method proposed by C. Lanczos in 1950 [67].

Specifically, given two Krylov subspaces $K_q(A, \mathbf{b})$, $K_q(A^T, \mathbf{c})$, the non-symmetric Lanczos method [32, 102] is as follows:

LANCZOS ALGORITHM $(A, \mathbf{b}, \mathbf{c})$
Compute $\rho_1 = \|\mathbf{b}\|_2$; $\eta_1 = \|\mathbf{c}\|_2$; $\mathbf{v}_1 = \frac{\mathbf{b}}{\rho_1}$; $\mathbf{w}_1 = \frac{\mathbf{c}}{\eta_1}$ —- # normalize **b** and **c**
Set $\mathbf{v}_0 = \mathbf{w}_0 = 0$ and $\tau_0 = 0$
For $j = 1, ..., q$ Do
Compute $\tau_j = \mathbf{w}_j^T \mathbf{v}_j$
Compute $\alpha_j = \frac{\mathbf{w}_j^T A \mathbf{v}_j}{\tau_j}$ $\beta_j = \frac{\tau_j}{\tau_{j-1}}\eta_j$; $\gamma_j = \frac{\tau_j}{\tau_{j-1}}\rho_j$
Compute $\mathbf{v} = A\mathbf{v}_j - \mathbf{v}_j\alpha_j - \mathbf{v}_{j-1}\beta_j$
Compute $\mathbf{w} = A^T\mathbf{w}_j - \mathbf{w}_j\alpha_j - \mathbf{w}_{j-1}\gamma_j$ —- # orthogonalization
If $\mathbf{v} = 0$ or $\mathbf{w} = 0$, stop, otherwise $\rho_{j+1} = \|\mathbf{v}_j\|_2$ and $\eta_{j+1} = \|\mathbf{w}\|_2$
$\mathbf{v}_{j+1} = \frac{\mathbf{w}}{\rho_{j+1}}$ and $\mathbf{w}_{j+1} = \frac{\mathbf{w}}{\eta_{j+1}}$ —- # normalization
EndDo

Figure 2.5 Non-symmetric Lanczos method for SISO systems.

After the Lanczos process, we have two matrices $V = \{\mathbf{v}_1, \mathbf{v}_2, ..., \mathbf{v}_q\}$ and $W = \{\mathbf{w}_1, \mathbf{w}_2, ..., \mathbf{w}_q\}$ and it can be shown that

$$W^T V = I, \tag{2.70}$$

where columns of W and V are the bases for Krylov subspaces $K_q(A, \mathbf{b})$, $K_q(A^T, \mathbf{c})$, respectively.

In addition to the two mutually orthogonal projection matrices, we also obtain two tri-diagonal matrices

$$T_q = \begin{bmatrix} \alpha_1 & \beta_2 & 0 & \cdots & 0 \\ \rho_2 & \alpha_2 & \beta_3 & \ddots & \vdots \\ 0 & \rho_3 & \ddots & \ddots & 0 \\ \vdots & \ddots & \ddots & \ddots & \beta_q \\ 0 & \cdots & 0 & \rho_q & \alpha_q \end{bmatrix}, \tag{2.71}$$

and

$$\tilde{T}_q = \begin{bmatrix} \alpha_1 & \gamma_2 & 0 & \cdots & 0 \\ \eta_2 & \alpha_2 & \gamma_3 & \ddots & \vdots \\ 0 & \eta_3 & \ddots & \ddots & 0 \\ \vdots & \ddots & \ddots & \ddots & \gamma_q \\ 0 & \cdots & 0 & \eta_q & \alpha_q \end{bmatrix}, \tag{2.72}$$

where

$$V^T A V = T_q, \tag{2.73}$$
$$W^T A^T W = \tilde{T}_q. \tag{2.74}$$

The Lanczos algorithm given in Figure 2.5 is just for the reduction of single-input and single-out system. For multi-input and multi-output systems, block Lanczos methods are needed [14, 33, 38].

2.4.5 RLC circuit formulation for reduction

For a general RLC linear network, we can apply modified nodal analysis to formulate it into the state space equations:

$$\begin{aligned} G\mathbf{x}(t) + C\dot{\mathbf{x}}(t) &= B\mathbf{u}(t) \\ \mathbf{y}(t) &= L^T\mathbf{x}(t), \end{aligned} \tag{2.75}$$

where G and C are the $n \times n$ conductive and storage element matrices, B and L are the $n \times N$ input and output positions matrices, and, typically, $B = L$ or $B = -L$; N is the number of input or output ports. State variables $\mathbf{x}$ can be nodal voltage or branch currents of the linear circuits. The matrices G and C can be further expressed as [14, 129]

$$G = \begin{bmatrix} A_g \mathcal{G} A_g^T & A_l & A_v \\ A_l^T & 0 & 0 \\ A_v^T & 0 & 0 \end{bmatrix} \quad C = \begin{bmatrix} A_c \mathcal{C} A_c^T & 0 & 0 \\ 0 & -\mathcal{L} & 0 \\ 0 & 0 & 0 \end{bmatrix} \quad \mathbf{x} = \begin{bmatrix} \mathbf{x}_n \\ \mathbf{x}_l \\ \mathbf{x}_v \end{bmatrix}, \tag{2.76}$$

where $\mathcal{G}$, $\mathcal{C}$, and $\mathcal{L}$ are the conductance, capacitance, and inductance sub-matrices. A_g, A_c, A_l and A_v are the incident matrices for conductive, capacitive, inductive, and voltage source elements [129].

The matrices $A_g \mathcal{G} A_g^T$, $A_c \mathcal{C} A_c^T$, and $\mathcal{L}$ are symmetric and positive semidefinite. As a result, G and C are symmetric and positive semidefinite.

We can further rewrite (2.76) into the following condensed form

$$G = \begin{bmatrix} G_{11} & G_{12} \\ G_{12}^T & 0 \end{bmatrix} \quad C = \begin{bmatrix} C_{11} & 0 \\ 0 & -C_{22} \end{bmatrix} \quad \mathbf{x} = \begin{bmatrix} \mathbf{x}_1 \\ \mathbf{x}_2 \end{bmatrix}, \tag{2.77}$$

where

$$G_{11} = A_g \mathcal{G} A_g^T \quad C_{11} = A_c \mathcal{C} A_c^T \quad C_{22} = \begin{bmatrix} \mathcal{L} & 0 \\ 0 & 0 \end{bmatrix}. \tag{2.78}$$

Notice that G_{11} and C_{11} are symmetric and positive semidefinite.

If we change the sign of G_{12}^T and the sign of $\mathbf{x}_2$, then we have

$$G = \begin{bmatrix} G_{11} & G_{12} \\ -G_{12}^T & 0 \end{bmatrix} \quad C = \begin{bmatrix} C_{11} & 0 \\ 0 & C_{22} \end{bmatrix} \quad \mathbf{x} = \begin{bmatrix} \mathbf{x}_1 \\ -\mathbf{x}_2 \end{bmatrix} \tag{2.79}$$

The resulting G and C become positive semidefinite, although they are no longer symmetric. This can easily be seen as

$$\begin{aligned} \mathbf{x}^T G \mathbf{x} &= \begin{bmatrix} \mathbf{x}_1^T & \mathbf{x}_2^T \end{bmatrix} \begin{bmatrix} G_{11} & G_{12} \\ -G_{12}^T & 0 \end{bmatrix} \begin{bmatrix} \mathbf{x}_1 \\ \mathbf{x}_2 \end{bmatrix} \\ &= \mathbf{x}_1^T G_{11} \mathbf{x}_1 + \mathbf{x}_1^T G_{12} \mathbf{x}_2 - \mathbf{x}_2^T G_{12}^T \mathbf{x}_1 \\ &= \mathbf{x}_1^T G_{11} \mathbf{x}_1 \geq 0, \end{aligned} \tag{2.80}$$

where $\mathbf{x}_1^T G_{11} \mathbf{x}_1 \geq 0$ means that $\mathbf{x}_1^T G_{11} \mathbf{x}_1$ is positive semidefinite. It will be shown in a later subsection that this property is important for passive model order reduction.

In the case of RC and RL circuits, if the circuits are excited by the current sources only (the transfer function is an admittance), the G and C and $\mathbf{x}$ become

$$G = A_g \mathcal{G} A_g^T, \quad C = A_c \mathcal{C} A_c^T, \quad \mathbf{x} = \mathbf{x_n}, \tag{2.81}$$

with $B = L$. As a result, G and C are symmetric *and* positive semidefinite.

2.4.6 *Passivity preservation*

In this subsection, we present some important results regarding the passive preservation in the Krylov subspace projection-based model order reduction.

Passivity is an important property of many physical systems. A passive network does not generate energy. If the reduced order model loses its passivity, it may lead to unbounded responses in transient simulation, which means that new energy has been generated in this network. Figure 2.6 shows a transient simulation result of a non-passive circuit under a sinusoidal excitation.

Brune [12] has proved that the admittance and impedance matrix of an electrical circuit consisting of an interconnection of a finite number of positive R, positive C, positive L, and transformers are passive if and only if their rational functions are positive real. A network with transfer matrix function $\mathbf{H}(s)$ is said to be positive real iff

$$\mathbf{H}(s) \text{ is analytic, for} Re(s) > 0. \tag{2.82}$$

$$\overline{\mathbf{H}}(s) = \mathbf{H}(\bar{s}) \text{ , for } Re(s) > 0, \tag{2.83}$$

$$\mathbf{H}_H(s) = \mathbf{H}(s) + \mathbf{H}(s)^H \geq 0 \text{ , for } Re(s) > 0, \tag{2.84}$$

where $\mathbf{H}_H(s)$ is called the Hermitian part of matrix $\mathbf{H}(s)$, as defined in (2.84). Mathematically, condition (2.82) means that there are no unstable poles (poles lie on the right-half-plane (RHP) in the s-domain). Condition (2.83) refers to a system that has a real response, and condition (2.84) is equivalent to stating that the real

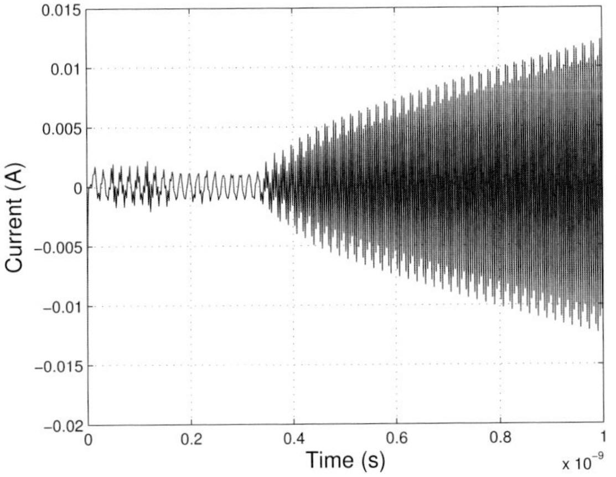

Figure 2.6 Transient response of a non-passive circuit.

part of $\mathbf{H}(s)$ is a positive semidefinite matrix at all frequencies. So it is only a necessary condition that a positive-real function has no poles in the RHP.

Conditions (2.82) and (2.83) are easily satisfied for the transfer functions of general RLC circuits when matrices G and C are real. Condition (2.84) is not satisfied in general. But if the matrices G and C are formulated in the passive form as shown in (2.79), then the transfer function is positive real and we have the following results [14, 85]:

Theorem 2.2. *For state space representation of an RLC system* $\{G, C, B\}$ *in* (2.75) *where* $B = L$. *If* $G + G^T \geq 0$ *and* $C = C^T \geq 0$ *and* $G + sC$ *is invertible at least at one point* s *with* $Re(s) > 0$, *then the following matrix transfer function is positive real*

$$H(s) = B^T(G + sC)^{-1}B. \tag{2.85}$$

The proof of the theorem can be found in [14]. The key step in the proof is the fact that the Hermitian of $W(s) = G + sC$, i.e., $W(s)_H = G + G^T + 2\delta C$ at $s = \delta + j\omega$ is positive semidefinite. The inverse of the matrix and congruence transformation also do not change the property of positive semidefinite.

After we perform the model order reduction on both G and C respectively, the reduced system is still passive. We have the following result [14]:

Theorem 2.3. *Let there be a state space representation of an RLC system* $\{G, C, B\}$ *in* (2.75) *where* $B = L$. *Let* V *be any* $n \times q$ *real reduction matrix. If* $G + G^T \geq 0$, $C = C^T \geq 0$ *and* V *is full rank. Let* $G_q = V^T G V$, $C_q = V^T C V$, *and*

$B_q = V^T B$. Then the matrix transfer function

$$H_q(s) = B_q^T (G_q + sC_q)^{-1} B_q \tag{2.86}$$

is positive real, thus passive.

2.4.7 PRIMA algorithm

In this section, we summarize the projection-based model order reduction by presenting the PRIMA (passive reduced-order interconnect macromodeling algorithm) algorithm [85], which performs general passive projection-based MOR on multi-input and multi-output (MIMO) RLC linear dynamic systems.

The PRIMA algorithm starts with the MNA formulation of a linear RLC dynamic MIMO system

$$\begin{aligned} G\mathbf{x}(t) + C\dot{\mathbf{x}}(t) &= B\mathbf{u}(t) \\ \mathbf{i}(t) &= B^T \mathbf{x}(t), \end{aligned} \tag{2.87}$$

where G and C are $n \times n$ conductance and storage element matrices in their passive form as shown in (2.79), $\mathbf{u}(t)$ and $\mathbf{i}(t)$ are the input port voltage and the current source vector at the same ports (to be computed) as outputs are measured at the inputs, B is the $n \times N$ position matrix for input and output ports, and N is the number of terminals (ports).

The transfer function, which is the admittance matrix, will become

$$Y(s) = B^T (G + sC)^{-1} B. \tag{2.88}$$

If we define

$$A = -G^{-1}C \quad R = G^{-1}B, \tag{2.89}$$

then the transfer function can be written as

$$Y(s) = B^T (I - sA)^{-1} R. \tag{2.90}$$

Assume that we have an $n \times q$ projection matrix V. Unlike some previous methods such as Pade via Lanczos (PVL) [32] or Block PVL (MPVL) [33, 38], which perform the reduction on the $I + A$ and R in (2.89), PRIMA performs the reduction on $G + sC$ and B directly. The major benefit of this reduction process is that $G + sC$ can be formed into a passive form as shown in Theorem 2.2. After the reduction by the congruence transformation, the resulting reduced models can be still passive as shown in Theorem 2.3.

Specifically, the reduced system becomes

$$\begin{aligned} V^T G V \mathbf{z}(t) + V^T C V \dot{\mathbf{z}}(t) &= V^T B \mathbf{u}(t) \\ \mathbf{i}(t) &= B^T V \mathbf{z}(t), \end{aligned} \tag{2.91}$$

where $\mathbf{z}$ is the reduced state vector of size q.

The reduced transfer function becomes

$$Y_q(s) = B_q^T (G_q + sC_q) B_q, \tag{2.92}$$

where

$$G_q = V^T G V \quad C_q = V^T C V \quad B_q = V^T B. \tag{2.93}$$

PRIMA applies a *block Arnoldi* algorithm to generate the projection matrix V. The reduced model matches the original systems in terms of *block moments* as defined below,

$$Y(s) = M_0 + M_1 s + M_2 s^2 + ... \tag{2.94}$$

where $M_i \in \mathcal{R}^{N\times N}$. The block moments can be computed as follows

$$M_i = B^T A^i R. \tag{2.95}$$

Accordingly, we can define a *block Krylov subspace* for a given matrix A and $R = [\mathbf{r}_0, \mathbf{r}_1 ..., \mathbf{r}_N]$

$$K_q(A, R) = \text{span}\{R, AR, ..., A^{k-1}R\}, \tag{2.96}$$

where $k = q/N$. If q/N does not result as in integer, we set $k = \lfloor q/N \rfloor$[2.1] and the corresponding Krylov subspace is defined as

$$K_q(A, R) = \text{span}\{R, AR, ..., A^{k-1}R, A^k\mathbf{r}_1, A^k\mathbf{r}_2, ..., A^k\mathbf{r}_l\}, \tag{2.97}$$

where $\mathbf{r}_i$ is the ith column vector of R, and $l = q - kN$.

With those concepts, given two matrices A and R, the block Arnoldi algorithm is presented as follows [14]:

BLOCK ARNOLDI ALGORITHM (A, R)

Compute QR factorization $R = V_0 X$
For $j - 1, ..., q/N$ Do
  Compute $V_j^0 = AV_{j-1}$
  For $i = 1, ..., j$ Do
    Compute $H_{j-i,j-1} = V_{j-i}^T V_j^{(i-1)}$
    Compute $V_j^{(i)} = V_j^{(i-1)} - V_{j-i} H_{j-i,j-1}$;
  EndDo
  Compute QR factorization $V_j^{(j)} = V_j H_{j,j-1}$
EndDo

Figure 2.7 Block Arnoldi method for MIMO systems.

After the Block Arnoldi process, we have the projection matrix $V = \{V_0, V_1, ...V_{k-1}\}$, whose columns of V form bases for the Krylov subspace generated by A and R, i.e.,

$$\text{colsp}(V) = K_q(A, R) \tag{2.98}$$

2.1 The $\lfloor . \rfloor$ operator is the truncation to the nearest integer towards zero.

and $V^TV = I$. The algorithm also generates a $q \times q$ block upper Hessenberg matrix

$$H_q = \begin{bmatrix} H_{00} & H_{01} & \cdots & H_{0,k-1} \\ H_{10} & H_{11} & \cdots & H_{1,k-1} \\ \vdots & \ddots & \ddots & \vdots \\ 0 & \cdots & H_{k-1,k-2} & H_{k-1,k-2} \end{bmatrix}, \tag{2.99}$$

which satifies

$$V^TAV = H_q. \tag{2.100}$$

2.5 Numerical examples

In this section, we present one numerical example for compact modeling by projection-based model order reduction. The example is a two-port large lumped RLC circuit. The values are marked in the Figure 2.8.

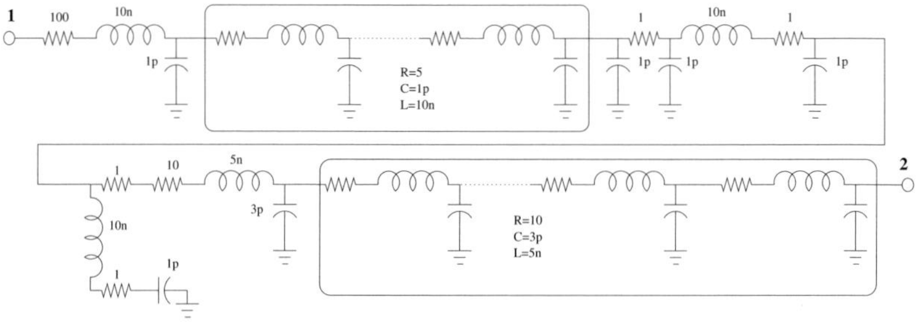

Figure 2.8 A two-port large lumped RCL circuit.

We perform projection-based model order reduction on the circuit with different orders (different poles). The results are shown in Figure 2.9 for $Y(11)$ responses and Figure 2.10 for $Y(12)$ responses. Hence, with more orders or more poles used, the reduced models will better match the original models.

2.6 Historical notes

In this section, we briefly review model order reduction methods based on Krylov subspaces from a historical perspective in the computer-aided design community.

Projection-based model order reduction techniques have been intensively studied in the past two decades [32, 37, 63, 85, 91, 113]. Projection-based methods were pioneered by asymptotic waveform evaluation (AWE) algorithm [91], where explicit moment matching was used to compute dominant poles at low frequency. The Pade via Lanczos (PVL) [32], block PVL (MPVL) [33], symmetric block PVM (SyMPVL) [38], and Arnoldi transformation [113] methods improved the numerical stability of AWE, while the split congruence transformation [63] method and

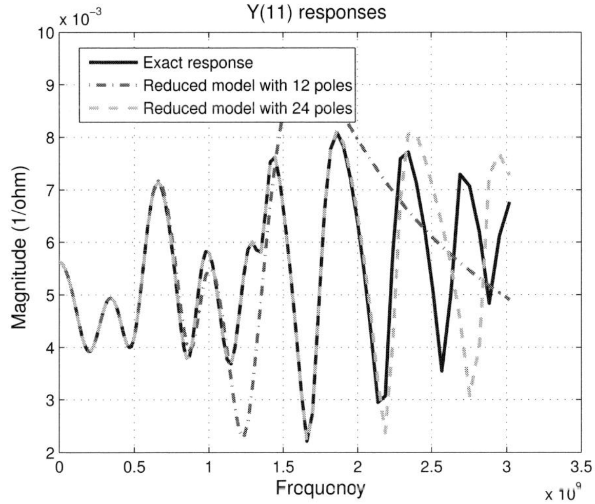

Figure 2.9 Comparison of the magnitudes of $Y(11)$ for different reduction orders for the lumped RLC circuit.

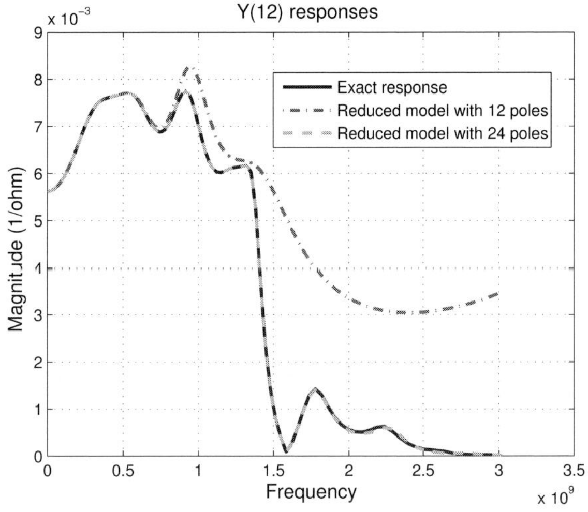

Figure 2.10 Comparison of the magnitudes of $Y(12)$ for different reduction orders for the lumped RLC circuit.

PRIMA [85] can further produce passive models. However, reduced circuit matrices by PRIMA are larger than direct pole marching (having more poles than necessary) [1] and PRIMA does not preserve certain important circuit properties such reciprocity [37]. The latest development by structured projection can preserve reciprocity [37], but it does not realize the reduced circuit matrices. The extension of structure-preserving MOR can be found in Chapter 8 of this book.

2.7 Summary

In this chapter, we presented the basic concepts for the model order reduction of interconnect circuits based on Krylov subspaces. We started with the concepts of moments, Elmore delays, and the Pade approximation for moment matching. We introduced the classic moment matching techniques like AWE, which suffer from the numerical issues. Then we presented the projection-based reduction framework and concepts of Krylov subspaces, which lead to more numerical stable model order reduction methods. Finally, we showed how model order reductions can be applied to RLC circuits and produce passive models via some numerical stable processes like Arnoldi and Lanczos methods for both SISO and MIMO systems.

2.8 Appendices

Theorem 2.1 (Moment-matching connection of one-side Krylov subspace). *For the two Krylov subspaces* colsp$V = K_q(A^{-1}E, A^{-1}\mathbf{b})$ *and* colsp$W = K_q(A^{-T}E^T, A^{-T}\mathbf{l})$, *defined in* (2.65). *For the one-side Krylov subspace method, the reduced system obtained by the projection-based reduction matches q moments of the transfer function of the original system defined in* (2.66).

Proof: We first rewrite the reduced system (2.57) as

$$E_r\dot{\mathbf{z}}(t) = A_r\mathbf{z}(t) + \mathbf{b}_r u(t)$$
$$y(t) = \mathbf{l}_r^T\mathbf{z}(t),$$

where $E_r = W^TEV$, $A_r = W^TAV$, $\mathbf{b}_r = W^T\mathbf{b}$, and $\mathbf{l}_r^T = \mathbf{l}^TV$. Then the reduced system transfer function can be written into

$$H_r(s) = -\mathbf{l}_r^TA_r^{-1}\mathbf{b}_r - \mathbf{l}_r^T(A_r^{-1}E_r)A_r^{-1}\mathbf{b}_r s, \cdots, -\mathbf{l}_r^T(A_r^{-1}E_r)^iA_r^{-1}\mathbf{b}_r s^i \cdots.$$

We first compare the moment 0 of the reduced system m_{r0} with the original system by using the Krylov subspace colsp$V = K_q(A^{-1}E, A^{-1}\mathbf{b})$.

$$\begin{aligned} m_{r0} &= \mathbf{l}_r^TA_r^{-1}\mathbf{b}_r \\ &= \mathbf{l}^TV(W^TAV)^{-1}W^T\mathbf{b} \\ &= \mathbf{l}^TV(W^TAV)^{-1}(W^TAV)\mathbf{r_0} \\ &= \mathbf{l}^TV\mathbf{r_0} = \mathbf{l}^TA^{-1}\mathbf{b} = m_0. \end{aligned}$$

The critical step in the above derivation is to realize that $A^{-1}\mathbf{b}$ belongs to the Krylov subspace $K_q(A^{-1}E, A^{-1}\mathbf{b})$ and, therefore, we can represent $A^{-1}\mathbf{b} = V\mathbf{r_0}$ and $\mathbf{b} = AV\mathbf{r_0}$.

We can also prove $m_{r0} = m_0$ by using Krylov subspace colsp$W =$

$K_q(A^{-T}E^T, A^{-T}\mathbf{l})$.

$$\begin{aligned} m_{r0} &= \mathbf{l}_r^T A_r^{-1}\mathbf{b}_r \\ &= \mathbf{l}^T V(W^T AV)^{-1}W^T\mathbf{b} \\ &= \mathbf{g_0}^T(W^T AV)(W^T AV)^{-1}W^T\mathbf{r_0} \\ &= \mathbf{g_0}^T W^T\mathbf{b} = \mathbf{l}^T A^{-1}\mathbf{b} = m_0. \end{aligned}$$

In this derivation, we utilize the fact that $A^{-T}\mathbf{l}$ belongs to the Krylov subspace $K_q(A^{-T}E^T, A^{-T}\mathbf{l})$ and, therefore, $A^{-T}\mathbf{l} = W\mathbf{g}_0$ and $\mathbf{l}^T = \mathbf{g}_0 W^T A$.

Then we compare the moment 1 of the reduced system m_{r1} and that of the original system, m_1. Still we first use Krylov subspace colsp$V = K_q(A^{-1}E, A^{-1}\mathbf{b})$ and we have

$$\begin{aligned} m_{r1} &= \mathbf{l}_r^T(A_r^{-1}E_r)A_r^{-1}\mathbf{b}_r \\ &= \mathbf{l}^T V\left[(W^T AV)^{-1}W^T EV\right](W^T AV)^{-1}W^T\mathbf{b} \\ &= \mathbf{l}^T V\left[(W^T AV)^{-1}W^T EV\right]\mathbf{r}_0 \\ &= \mathbf{l}^T V(W^T AV)^{-1}W^T A(A^{-1}E)A^{-1}\mathbf{b} \\ &= \mathbf{l}^T V(W^T AV)^{-1}(W^T AV)\mathbf{r}_1 \\ &= \mathbf{l}^T V\mathbf{r}_1 = \mathbf{l}^T(A^{-1}E)A^{-1}\mathbf{b} = m_1. \end{aligned}$$

where $V\mathbf{r}_1 = (A^{-1}E)A^{-1}b$ as $(A^{-1}E)A^{-1}b$ belongs to $K_q(A^{-1}E, A^{-1}\mathbf{b})$.

In general, we have $(A^{-1}E)^i A^{-1}\mathbf{b} = V\mathbf{r}_i, i < q$ for the Krylov subspace $K_q(A^{-1}E, A^{-1}\mathbf{b})$. Then we can prove that

$$\begin{aligned} m_{ri} &= \mathbf{l}_c^T(A_r^{-1}E_r)^i A_r^{-1}\mathbf{b}_r \\ &= \mathbf{l}^T V\left[(W^T AV)^{-1}W^T EV\right]^i(W^T AV)^{-1}W^T\mathbf{b} \\ &= \mathbf{l}^T V\mathbf{r}_i = \mathbf{l}^T(A^{-1}E)^i A^{-1}\mathbf{b} = m_i. \end{aligned}$$

Similarly, if we use Krylov subspace $W = K_q(A^{-T}E^T, A^{-T}\mathbf{l}), i < q$. As a result, we can have $(A^{-T}E^T)^i A^{-T}\mathbf{l} = W\mathbf{g}_i$. We then show that

$$\begin{aligned} m_{ri} &= \mathbf{l}_c^T(A_r^{-1}E_r)^i A_r^{-1}\mathbf{b}_r \\ &= \mathbf{l}^T V\left[(W^T AV)^{-1}W^T EV\right]^i(W^T AV)^{-1}W^T\mathbf{b} \\ &= \mathbf{g}_i^T W^T\mathbf{b} = \mathbf{l}^T A^{-1}(EA^{-1})^i\mathbf{b} \\ &= \mathbf{l}^T(A^{-1}E)^i A^{-1}\mathbf{b} = m_i. \end{aligned}$$

QED.

Corollary 2.1 (Moment-matching connection of two-side Krylov subspace methods). *If both subspaces V and W are used, the reduced model matches $2q$ moments of the transfer function of the original system defined in* (2.66).

Proof: To prove the corollary, we only need to look at the expected highest-order moment of the reduced system m_{2q-1};

$$\begin{aligned}
m_{r2q-1} &= \mathbf{l}_c^T(A_r^{-1}E_r)^{2q-1}A_r^{-1}\mathbf{b}_r \\
&= \mathbf{l}_c^T(A_r^{-1}E_r)^{q-1}(A_r^{-1}E_r)(A_r^{-1}E_r)^{q-1}A_r^{-1}\mathbf{b}_r \\
&= \mathbf{l}^T V\left[(W^TAV)^{-1}W^TEV\right]^{q-1} \times \\
&\quad \left[(W^TAV)^{-1}W^TEV\right]\left[(W^TAV)^{-1}W^TEV\right]^{q-1}(W^TAV)^{-1}W^T\mathbf{b} \\
&= \mathbf{g}_{q-1}^T(W^TAV)\left[(W^TAV)^{-1}W^TEV\right]\mathbf{r}_{q-1} \\
&= \mathbf{g}_{q-1}^T W^TEV\mathbf{r}_{q-1} \\
&= \mathbf{l}^TA^{-1}(EA^{-1})^{q-1}E(A^{-1}E)^{q-1}A^{-1}\mathbf{b} \\
&= \mathbf{l}^T(A^{-1}E)^{q-1}A^{-1}E(A^{-1}E)^{q-1}A^{-1}\mathbf{b} \\
&= \mathbf{l}^T(A^{-1}E)^{2q-1}A^{-1}\mathbf{b} = m_{2q-1}.
\end{aligned}$$

QED.

3 Truncated balanced realization methods for MOR

3.1 Introduction

Model order reduction methods for linear and non-linear dynamic systems in general can be classified into two categories [6]:

1. Singular-value-decomposition (SVD) based approaches
2. Krylov-subspace-based approaches.

Krylov-subspace-based methods have been reviewed in Chapter 2. In this chapter, we focus on the SVD-based reduction methods. Singular value decomposition is based on the lower rank approximation, which is optimal in the 2-norm sense. The quantities for deciding how a given system can be approximated by a lower-rank system are called *singular values*, which are the square roots of the eigenvalues of the product of the system matrix and its adjoint. The major advantage of SVD-based approaches over Krylov subspace methods lies in their ability to ensure the errors satisfying an a-priori upper bound. Also, SVD-based methods typically lead to optimal or near optimal reduction results as the errors are controlled in a global way. However, SVD-based methods suffer the scalability issue as SVD is a computational intensive process and cannot deal with very large dynamic systems in general. In contrast, Krylov-subspace-based methods can scale to reduce vary large systems due to efficient computation methods for moment vectors and their orthogonal forms.

SVD-based approaches consist of several reduction methods [6]. In this chapter, we mainly focus on the *truncated-balanced-realization (TBR)* approach and its variants, which were first introduced by Moore [81]. In the TBR method, a system is mapped onto a basis where the states that are difficult to reach are also difficult to observe. Then, the reduced model is obtained simply by truncating those weak states.

TBR-based approaches have been applied to reduce linear VLSI systems in the past [87,89,131]. Also, TBR methods can produce passive models by using positive real TBR methods [87,89,131]. The latest development by using empirical Gramians improves the scalability of the TBR methods [89,90].

Here, we first review the singular value decomposition method in Section 3.2. Then, we briefly discuss the proper orthogonal decomposition (POD) method, as it can be viewed as a simple application of SVD on the responses of a dynamic system in section 3.3. After this we proceed to the TBR-based reduction methods in the remaining sections.

3.2 The singular value decomposition (SVD)

In this section, we briefly review the singular value decomposition method, as it is the key reduction step used in this chapter.

For an $m \times n$ matrix A, the SVD decomposition of A is

$$A = U_{m\times m} \Sigma V_{n\times n}^T, \tag{3.1}$$

where $U_{m\times m}$ and $V_{n\times n}$ are orthogonal matrices, $U_{m\times m}^T U_{m\times m} = I$ and $V_{n\times n}^T V_{n\times n} = I$. $\Sigma = \mathrm{diag}(\sigma_1, \sigma_2, \ldots, \sigma_{\min(m,n)})$, σ_i is called singular values: $\sigma_i = \sqrt{\lambda_i(A^T A)} \geq \sigma_{i+1}$ and $\sigma_1 \geq \sigma_2 \geq \cdots \geq \sigma_{\min(m,n)}$.

SVD can lead to the best approximation in terms of the 2-norm [44], which is expressed by the following theorem:

Theorem 3.1 (Schmidt–Mirsky, Eckart–Young). *For an SVD decomposition of a matrix A, if $k < r = \mathrm{rank}(A)$ and*

$$A_k = \sum_{i=1}^{k} \sigma_i u_i v_i^T, \tag{3.2}$$

then

$$\min_{\mathrm{rank}(B)=k} = \| A - A_k \|_2 = \sigma_{k+1}, \tag{3.3}$$

where $\{u_i\}$ and $\{v_i\}$ are the left and right singular vectors respectively.

Equation (3.3) reflects the fact that the rank-k approximation matrix A_k is just σ_{k+1} away from original matrix A in terms of the 2-norm distance. Hence, SVD gives a very good low-rank approximation to the original system with errors bounded by the known singular values.

3.3 Proper orthogonal decomposition (POD)

We first start with the proper orthogonal decomposition method as it is a straightforward way to apply the SVD method for model order reduction. This method can be used for both linear and non-linear system reductions.

Consider the dynamic (linear and non-linear) system given by

$$\dot{x}(t) = f(x(t), u(t)), \tag{3.4}$$

$$y(t) = h(x(t), u(t)), \tag{3.5}$$

where $u(t)$ is the input vector the system. For a fixed input $u(t)$, the state trajectory at certain instances of time (m samples) is measured:

$$\mathcal{X} = [x(t_1), x(t_2), ..., x(t_m)]. \tag{3.6}$$

the $\mathcal{X}$ matrix can be viewed as the snapshots of the state. In general $m \gg n$. Next we perform SVD on $\mathcal{X}$. If the singular values of this matrix fall off rapidly, a low-order approximation of this system can be computed,

$$\mathcal{X} = U\Sigma V^T \approx U_k \Sigma_k V_k^T, \tag{3.7}$$

where Σ and Σ_k are diagonal matrices with singular values in a decreasing order along the diagonal and Σ_k is a $k \times k$ diagonal matrix. As a result, we have the *eigendecomposition* of the response Gramian as

$$\mathcal{X}\mathcal{X}^T = U\Sigma\Sigma^T U^T \approx U_k \Sigma_k \Sigma_k^T U_k^T. \tag{3.8}$$

Let $x(t) \approx U_k \eta(t), \eta(t) \in \mathcal{R}^k$; then we have the following reduced system in terms of $\eta(t)$,

$$\dot{\eta}(t) = U_k T f(U_k \eta(t), u(t)), \tag{3.9}$$
$$y(t) = h(U_k \eta(t), u(t)). \tag{3.10}$$

Thus the approximation of the state $x(t)$ uses a low-dimensional space that is spanned by the k leading columns of U, which essentially are the dominant eigen vectors of the response Gramian $\mathcal{X}\mathcal{X}^T$. This reduction process can be used directly for both linear and non-linear dynamic systems. But the state trajectory is dependent on the input signals.

The POD method is more suitable for autonomous systems like oscillators as we perform the decomposition directly on the state trajectory of the system. For systems with input and output terminals, it is more desirable to perform the reduction on the controllability and observability Gramians as they are better ways to describe the system dynamic behaviors from the input's and output's perspectives. This leads to the truncated balanced realization (TBR) methods, to be discussed in the following sections.

3.4 Classic truncated balanced realization methods

In this section, we review the classic TBR method for general linear dynamic systems.

3.4.1 State-space models

Given a linear system in its state-space form,

$$\frac{\mathrm{d}x}{\mathrm{d}t} = Ax(t) + Bu(t), \tag{3.11}$$

$$y(t) = Cx(t) + Du(t), \tag{3.12}$$

where $A \in R^{n\times n}$, $B \in R^{n\times p}$, $C \in R^{p\times n}$, $D \in R^{p\times p}$, $y(t)$, and $u(t) \in R^p$, model reduction algorithms seek to produce a similar system

$$\frac{\mathrm{d}\tilde{x}}{\mathrm{d}t} = \tilde{A}\tilde{x}(t) + \tilde{B}u(t), \tag{3.13}$$

$$\tilde{y}(t) = \tilde{C}\tilde{x}(t) + \tilde{D}u(t), \tag{3.14}$$

where $\tilde{A} \in R^{q\times q}$, $\tilde{B} \in R^{q\times p}$, $\tilde{C} \in R^{p\times q}$, and $\tilde{D} \in R^{p\times p}$, of order q much smaller than the original order n, but for which the outputs $y(t)$ and $\tilde{y}(t)$ are approximately equal for inputs $u(t)$ of interest. Often the transfer functions

$$H(s) = D + C(sI - A)^{-1}B, \tag{3.15}$$

$$\tilde{H}(s) = \tilde{D} + \tilde{C}(sI - \tilde{A})^{-1}\tilde{B}, \tag{3.16}$$

are used as a metric for approximation: if $\|H(s) - \tilde{H}(s)\| < \epsilon$, in some appropriate norm, for some given allowable error ϵ and allowed domain of the complex frequency variable s, the reduced model is accepted as accurate.

3.4.2 Basic idea of balanced truncation

Consider a system as in (3.11) where A is stable or, equivalently, its spectrum is in the open left half plane.

We consider the output response to a particular input signal, namely, the impulse $u = \delta$. In this case the output is called the impulse response and is denoted by $h(t) = Ce^{At}B, t \geq 0$. This can be decomposed into an input-to-state map $x(t) = e^{At}B$, and a state-to-output map $\eta(t) = Ce^{At}$. Thus the input δ causes the state $x(t)$, while the initial condition $x(0)$ causes the output $y(t) = \eta(t)x(0)$. The Gramians corresponding to x and η are

$$W_c = \sum_t x(t)x(t)^T = \int_0^\infty \mathrm{e}^{At}BB^T\mathrm{e}^{A^Tt}\mathrm{d}t, \tag{3.17}$$

$$W_o = \sum_t \eta(t)^T\eta(t) = \int_0^\infty \mathrm{e}^{A^Tt}C^TC\mathrm{e}^{At}\mathrm{d}t. \tag{3.18}$$

where W_c and W_o are called the controllability and observability Gramians respectively.

A linear system in the state space form is called balanced if the solutions of the two Gramians are equal and diagonal:

$$W_c = W_o = \Sigma = \mathrm{diag}(\sigma_1, \sigma_2, \ldots, \sigma_n). \tag{3.19}$$

It turns out that every controllable and observable system can be transformed to the so-called balanced form by means of a basis change $\tilde{x} = Tx$. The two Gramians are transformed by congruence transformation:

$$\tilde{W}_c = TW_cT^T, \tag{3.20}$$

$$\tilde{W}_o = T^{-T}W_oT^{-1}, \tag{3.21}$$

$$\tilde{W}_c\tilde{W}_o = TW_cW_oT^{-1}. \tag{3.22}$$

Since any two symmetric matrices can be simultaneously diagonalized by an appropriate congruence transformation, it is possible to find a value for T that makes the $\tilde{W}_c$ and $\tilde{W}_o$ equal and diagonal. Since the product W_cW_o transforms under similarity at the same time, the eigenvalues of the product W_cW_o are invariant. These eigenvalues, $\lambda_i(W_cW_o)$, are called the *Hankel singular values*, and contain useful information about the input-output behavior of the system.

In particular, small eigenvalues of W_cW_o correspond to internal sub-systems that have a weak effect on the input-output behavior of the system and are almost non-observable or non-controllable or both. These are fundamental invariants that determine how well a model can be approximated by a reduced-order model. They play the same role for dynamical systems that the singular values play for finite-dimensional matrices.

The key for the computation of the Hankel singular values is the observation that the Gramians W_c, W_o are the unique Hermitian positive definite solutions to the following linear matrix equations, which are known as Lyapunov equations:

$$AW_c + W_cA^T + BB^T = 0, \tag{3.23}$$

$$A^TW_o + W_oA + C^TC = 0. \tag{3.24}$$

Then, the Hankel singular values of the system are the square roots of the eigenvalues of the product W_cW_o.

If a system is controllable and observable, the two Gramians (Wc and Wo) are positive definite and then can be factored as $Wc = L_cL_c^T$ and $Wo = L_oL_o^T$ respectively by Cholesky factorization, where L_c and L_o are lower triangular matrices. Let $L_o^TL_c = U\Sigma V^T$ be the SVD of $L_o^TL_c$. The balancing transformation matrix can be obtained as

$$T = L_cV\Sigma^{-\frac{1}{2}}, \tag{3.25}$$

$$T^{-1} = \Sigma^{-\frac{1}{2}}U^TL_o^T. \tag{3.26}$$

Then, under a similarity transformation of the state-space model,

$$\tilde{A} = T^{-1}AT, \tag{3.27}$$

$$\tilde{B} = T^{-1}B, \tag{3.28}$$

$$\tilde{C} = CT. \tag{3.29}$$

We may partition Σ into

$$\Sigma = \begin{bmatrix} \Sigma_1 & 0 \\ 0 & \Sigma_2 \end{bmatrix}. \tag{3.30}$$

Conformally partitioning the transformed matrices as

$$\tilde{A} = \begin{bmatrix} \tilde{A}_{11} & \tilde{A}_{12} \\ \tilde{A}_{21} & \tilde{A}_{22} \end{bmatrix}, \tag{3.31}$$

$$\tilde{B} = \begin{bmatrix} \tilde{B}_1 \\ \tilde{B}_2 \end{bmatrix}, \tag{3.32}$$

$$\tilde{C} = \begin{bmatrix} \tilde{C}_1 & \tilde{C}_2 \end{bmatrix}. \tag{3.33}$$

The reduced-order model is obtained by simple truncation, that is, by taking the $k \times k$, $k \times m$, $p \times k$ leading blocks of $\tilde{A}$,$\tilde{B}$, $\tilde{C}$, respectively; the system satisfies kth order Lyapunov equations with diagonal solution Σ_1. This truncation leads to a balanced reduced-order system (A_{11}, B_1, C_1, D).

Algorithm 3.1: CLASSIC TRUNCATED BALANCED REALIZATION(TBR)

1. Solve $AW_c + W_cA^T + BB^T = 0$ for W_c.
2. Solve $A^TW_o + W_oA + C^TC = 0$ for W_o.
3. Compute Cholesky factors, $W_c = L_cL_c^T$, $W_o = L_oL_o^T$.
4. Compute SVD of Cholesky factors $U\Sigma V^T = L_o^TL_c$ where Σ is diagonal positive and U,V have orthonormal columns.
5. Compute the balancing transformation matrices $T = L_cV\Sigma^{-1/2}$, $T^{-1} = \Sigma^{-1/2}U^TL_o^T$.
6. Form the balanced realization transformations as $\tilde{A} = T^{-1}AT$, $\tilde{B} = T^{-1}B$, $\tilde{C} = CT$.
7. Select reduced model order and partition realization $\tilde{A}$, $\tilde{B}$, $\tilde{C}$ conformally.
8. Truncate $\tilde{A}$, $\tilde{B}$, $\tilde{C}$ to form the reduced realization.

3.4.3 *Error bounds*

One of the attractive aspects of TBR methods is that computable error bounds are available. If this truncation is such that the resulting Gramians contain the largest k singular values σ_1 through σ_k, then the H_∞ norm of the error system has the following upper and lower bound [29, 43]:

$$\sigma_k \le \|H(s) - \tilde{H}_q(s)\|_\infty \le 2 \sum_{k=q+1}^{N} \sigma_k. \tag{3.34}$$

The H_∞ norm of a system Σ is defined as the maximum of the highest peak of the frequency response, i.e., as the largest singular value of the transfer function evaluated on the imaginary axis (i.e., of the frequency response): $\sigma_{\max}[D + C(j\omega - A)^{-1}B]$. This result provides rigorous justification for using the reduced-order model to predict behavior and enact control of the full system. The upper bound for the error system given above translates to an upper bound for the energy of the output error signal.

3.5 Passive-preserving truncated balanced realization methods

3.5.1 *Passivity and positive-realness*

A passive system cannot produce energy internally. It is desired that reduced models also be passive. Otherwise, the reduced models may cause non-physical behavior when used in system simulations, such as by generating energy at high frequencies that cause erratic or unstable time-domain behavior. For many electrical systems of interest, passivity is implied by positive-realness of the transfer function. The matrix $H(s)$ is said to be positive real (PR) if:

$$\overline{H}(s) = H(\overline{s}), \tag{3.35}$$

$$H(s) \text{ is analytic in } \{s : Re(s) > 0\}, \tag{3.36}$$

$$H(s) + H(s)^H \geq 0 \{s : Re(s) > 0\}. \tag{3.37}$$

In the above, $\overline{H}$ denotes complex conjugate, H^H denotes Hermitian (complex conjugate and transpose), and $\leq$ in a matrix context denotes semidefiniteness.

3.5.2 *Positive-realness constraints*

The central tool in relating passivity of state-space models to positive-realness of the transfer function is the positive-real lemma [4]. Let (A, B, C, D) describe a state-space model. The positive-real lemma states that the system is passive if and only if there exists a $P(\in R^{n \times n}) \geq 0$ satisfying the linear matrix inequality

$$\begin{bmatrix} A^T P + PA & PB - C^T \\ B^T P - C & -(D + D^T) \end{bmatrix} \leq 0. \tag{3.38}$$

Two important methods for generating guaranteed passive models of reduced systems, positive-real TBR [87] and convex optimization for passivity enforcement method [22], are both based on this lemma. We will introduce positive-real TBR (PR-TBR) in this section.

3.5.3 *Lur'e equations*

We assume $D + D^T > 0$. For RLC models in a modified nodal analysis (MNA) format, we have $A + A^T \leq 0$, $B = C^T$, and $D = 0$. The MNA system can then be transformed into an equivalent form with $D + D^T > 0$. We will discuss this case in the following sections.

Using the Schur complement, the solution of the positive-real lemma is equivalent to the following Lur'e equations. Therefore, the positive-real lemma means that $H(s)$ is positive real if and only if there exist matrices $X_c = X_c^T$, J_c, K_c such that the following Lur'e equations [5] are satisfied

$$AX_c + X_c A^T = -K_c K_c^T, \tag{3.39}$$

$$X_c C^T - B = -K_c J_c^T, \tag{3.40}$$

$$J_c J_c^T = D + D^T, \tag{3.41}$$

and $X_c \geq 0$ (X_c is positive semidefinite), X_c is analogous to the controllability Gramians. In fact, it is the controllability Gramian for a system with the input-to-state mapping given by the matrix K_c (K_c is treated as input matrix).

Similarly there is a dual set of Lur'e equations for $X_o = X_o^T > 0$: J_o , K_o are obtained from above equations by the substitutions $A \to A^T$, $B \to C^T$, $C^T \to B$. The dual equations

$$A^T X_o + X_o A = -K_o^T K_o, \tag{3.42}$$

$$X_o B - C^T = -K_o^T J_o, \tag{3.43}$$

$$J_o^T J_o = D + D^T, \tag{3.44}$$

have a corresponding observability quantity $X_o \geq 0$ for a positive-real $H(s)$. It can be verified that $X_c X_o$ transform under similarity just as $W_c W_o$, so that their eigenvalues are invariant, and in fact they behave as the Gramians W_c and W_o.

3.5.4 Algebraic Riccati equations

Combining the three equations gives

$$AX_c + X_c A^T + (B - X_c C^T)(D + D^T)(B - X_c C^T)^T = 0, \tag{3.45}$$

which is the so-called algebraic Riccati equation (ARE) [7]. Taking the matrix root $LL^T = (D + D^T)^{-1}$ and defining $B_0 = BL$, $C_0 = L^T C$, and $A_0 = A - B_0 C_0$, the ARE is rewritten as

$$A_0^T X_c + X_c A_0 + X_c B_0 B_0^T X_c + C_0^T C_0 = 0. \tag{3.46}$$

Also, there is a dual set of ARE that are obtained from above equations by the substitutions $A_0 \to A_0^T$, $B_0 \to C_0^T$, $C_0^T \to B_0$. The dual equation is

$$A_0 X_o + X_o A_0^T + X_o B_0 B_0^T X_o + C_0^T C_0 = 0. \tag{3.47}$$

3.5.5 Passivity-preserving truncated balanced realization

A passivity-preserving reduction procedure is to solve Lur'e equations or algebraic Riccati equations (AREs) for the quantities X_c,X_o, which may then be used as the basis of a TBR procedure.

We may find a coordinate system in which $\tilde{X}_c = \tilde{X}_o = \Sigma$, with Σ again being diagonal. In this coordinate system, the matrices $\tilde{A}, \tilde{B}, \tilde{C}$ may be partitioned and truncated, similar to the standard TBR procedure. Here is the outline of this algorithm [87].

Algorithm 3.2: PASSIVE TRUNCATED BALANCED REALIZATION

1. Solve Lur'e equations or AREs for X_c and X_o.
2. Proceed with steps 3 to 8 in Algorithm 3.1, substituting X_c for W_c and X_o for W_o.

3.6 Hybrid TBR and combined TBR-Krylov subspace methods

A hybrid TBR method was proposed in [87] where the classic TBR method is tried first. If the reduced models are not passive, then the positive-real TBR is tried. The rationale for doing this is that TBR approximates are more accurate for a given order than PR-TBR. Also PR-TBR costs more than TBR in computation. The hybrid TBR is presented as follows:

Algorithm 3.3: HYBRID TRUNCATED BALANCED REALIZATION

1. Perform Algorithm 3.2.
2. Using the reduced model matrices $\tilde{A}, \tilde{B}, \tilde{C}$, solve Lur'e equatons or AREs for $\tilde{X}_c$.
3. If those equations are solvable and $\tilde{X}_c \geq 0$, then terminate and return $\tilde{A}, \tilde{B}, \tilde{C}$; otherwise discard the TBR-reduced model and proceed with Algorithm 3.2.

Truncated balanced realization algorithms are important from the theoretical point of view. For small systems (a few hundred states or so) they are superior in accuracy to the Krylov subspace methods and other parameter-matching techniques, and also provide computable bounds on the reduction errors. For large systems, direct application of the techniques used to balance and truncate the systems is computationally infeasible, since the computations required have $O(n^3)$ complexity when performed directly (n being the order of the system to be reduced). Therefore, the TBR methods are of more interest when combined with iterative Krylov-subspace procedures.

The idea is to obtain an initial reduced model via some initial reduction or approximation technique and then further compress it using a TBR method. As a matter of fact, the initial approximation can be generated by any desired method, for example rational fitting, or a Krylov-subspace technique. Actually a combined TBR and Krylov approach was investigated in [60] for interconnect and packing modeling and nearly optimal reduced models were obtained using such a combined reduction scheme.

3.7 Empirical TBR and poor man's TBR

As mentioned already, the TBR method is a computationally expensive reduction process owing to the cubical time complexity for computing the Gramians. One strategy to overcome this problem is to use approximate Gramians, which can be computed more cheaply. Such a strategy is generally referred to as *Empirical TBR* and has been explored in the poor Man's TBR method for interconnect model order reduction [89,90]. Poor man's TBR also can be viewed as a special form of proper orthogonal decomposition methods applied to linear dynamic systems as shown in (3.11).

3.7.1 *Poor man's TBR*

In this subsection, we briefly review the poor man's TBR method [90]. The basic idea of Poor Man's TBR method is to recognize that the Gramian in the frequency-domain form can be approximated by using the state responses under impulse responses in frequency domain.

Specifically, given a linear system in its state-space form

$$\frac{\mathrm{d}x}{\mathrm{d}t} = Ax(t) + Bu(t), \tag{3.48}$$

$$y(t) = Cx(t), \tag{3.49}$$

where $A \in R^{n\times n}$, $B \in R^{n\times p}$, $C \in R^{p\times n}$, $y(t)$, and $u(t) \in R^p$. We further assume that a stable circuit under discussion has a symmetric state matrix $A = A^T$ and the same input and output terminals, i.e., $C = B^T$. In this case, both Gramians are equal. As a result, we only need to look at the controllability Gramian, X,

$$X = \int_0^{\infty} \mathrm{e}^{At} BB^T \mathrm{e}^{A^T t} \mathrm{d}t. \tag{3.50}$$

Since the Laplace transform of e^{At} is $(sI - A)^{-1}$, the Gramian X in the frequency domain becomes:

$$X = \int_{-\infty}^{\infty} (j\omega I - A)^{-1} BB^T (j\omega I - A)^{-H} \mathrm{d}\omega, \tag{3.51}$$

where superscript H denotes the Hermitian transpose. We now consider evaluating X by applying numerical quadrature to the above equation, given a quadrature scheme with nodes ω_k and weights ω_k, and defining

$$z_k = (j\omega_k I - A)^{-1} B. \tag{3.52}$$

Notice that (3.52) can essentially be viewed as the state response of (3.48) under impulse responses. For general POD methods, we can apply typical inputs to the system and obtain the responses. An approximation of $\hat{X}$ to X can then be computed as

$$\hat{X} = \sum_k w_k z_k z_k^H, \tag{3.53}$$

which can be viewed as the response matrices' Gramian.

Let Z be a matrix whose columns are z_k , and W a diagonal matrix with diagonal entries $W_{kk} = \sqrt{w_k}$. The above equation can be written more compactly as

$$\hat{X} = ZW^2Z^H = (ZW)(WZ^H) = (ZW)(ZW)^H. \tag{3.54}$$

Following the POD method, we perform singular value decomposition of ZW to perform the model reduction.

$$ZW = VSU, \tag{3.55}$$

with S_Z real diagonal, V and U unitary matrices. As expected, we have the eigen-decomposition of the approximate Gramian $\hat{X}$

$$\begin{aligned}\hat{X} &= (ZW)(ZW)^T \\ &= (VSU)(VSU)^T \\ &= VSUU^T SV^T \\ &= VS^2V^T.\end{aligned} \tag{3.56}$$

Hence, the dominant singular vectors in V are the eigenvectors of $\hat{X}$. Therefore, V converges to the eigenspaces of X, and the Hankel singular values are obtained directly from the entries of S. V can then be used as the projection matrix in a model order reduction scheme. The poor man's TBR method is illustrated in the following:

Algorithm 3.4: PMTBR: POOR MAN'S TBR

1. Do until satisfied:
2. Select a frequency point s_i.
3. Compute $z_i = (s_i I - A)^{-1}B$.
4. Form the matrix with columns: $Z_{(i)} = [z_1, z_2, \cdots, z_i]$.
5. Construct the SVD of $Z_{(i)}$. If the error is satisfactory, set $Z = Z_{(i)}$, go to Step 6. Otherwise, go to Step 2.
6. Construct the projection space V from the orthogonalized column span of Z, dropping columns whose associated singular values fall below a desired tolerance.

Because the full Gramians are not computed, as developed here, poor-man's TBR does not possess the stability and passivity preserving properties of the full-blown TBR algorithms [90].

3.8 Computational complexities of TBR methods

In this section, we briefly discuss the time complexities of TBR methods as this is important to understand the advantages and limitations of various TBR methods. Specifically, we analyze the time complexities of standard TBR and poor man's TBR methods. Similar analysis can be found at [90].

3.8.1 *Standard TBR methods*

We assume that we start with a linear dynamic system with n states, as in (3.48). We want to derive the cost of computing a qth order model and $q \ll n$. To analyze the TBR and its variants, we need to quote the time complexities of the basic operations involved.

For the SVD operations, the cost is roughly $O(nq^2)$ for $q \ll n$ [44]. Note that the time complexity of the full-blown SVD on an $n \times m$ matrix, where $n > m$, is approximately $O(nm^2)$ where all the singular values are computed. But, practically, we can compute the dominant q singular values instead, which can lead to $O(nq^2)$ complexity. The linear matrix solving takes $O(n^\alpha)$ (typically, $1 \leq \alpha \leq 1.2$ for sparse circuits), and matrix factorizations take $O(n^\beta)$ (typically, $1.1 \leq \beta \leq 1.5$ for sparse circuits). The formation of transformation matrices is $O(q^2 n)$ and similarity transformation takes roughly $O(nq)$.

The TBR algorithm implies the computation of the Gramians X and W, and the eigenvalues of their product, at a cost of $O(n^3)$. Since we need to take two Cholesky decompositions, with $O(n^\beta)$, the total cost of standard TBR is approximately

$$O(n^3 + 2n^\beta + nq^2 + nq). \tag{3.57}$$

Clearly, the $O(n^3)$ limits the TBR method to reduce large systems.

3.8.2 Poor man's TBR methods

Assuming that q frequency points are chosen in the quadrature scheme for PMTBR, an examination of Algorithm 3.4, indicates that it involves one SVD, q linear solutions, and q factorizations, for a total cost of

$$O(nq^2 + qn^\alpha + qn^\beta + nq^2 + nq). \tag{3.58}$$

The last two items are the cost of the generating the transformation matrix and performing the reductions as mentioned before.

As we can see, the poor man's TBR is much more scalable. It was shown that the computing cost of poor man's TBR is similar to that of the multi-point Krylov subspace methods [90].

3.9 Practical implementation and numerical issues

In this section, we discuss some practical implementation and numerical issues for the standard TBR algorithm and its variants.

3.9.1 Model reduction of unstable systems

Directly applying the TBR method mentioned above may encounter some numerical issues. One problem we found in practical implementation of TBR methods is that the controllability and observability Gramians computed from Lyapunov equations (3.24) may not be positive definite. As a result, we cannot perform the Cholesky decomposition in step (3) in Algorithm 3.1 and the algorithm simply breaks. The typical reason for this problem is the numerical noise in the reduction process, which can make a given system unstable even though the original system is stable and passive.

In this section, we present a method to deal with this problem [10]. The basic idea is to decompose the system into two parts, the stable part and unstable part, and then perform the reduction on the stable part only.

Given a system transfer function matrix $G(s)$, which can be rewritten into the following form:

$$G(s) = G_{-}(s) + G_{+}(s), \tag{3.59}$$

such that G_{-} is stable and G_{+} is unstable. Then any of the absolute or relative error state-space truncation methods for model reduction can be applied to G_{-} in order to obtain a reduced-order transfer function $\tilde{G}_{-}$, and the reduced order system is synthesized by $\tilde{G}(s) = \tilde{G}_{-}(s) + G_{+}(s)$.

For model order reduction of RLC circuits, which are stable and passive, we can just simply ignore the unstable part, $G_{+}(s)$, as it is definitely not a dominant part of the system. But for some control application, those unstable systems may be created specifically. Hence, we still need to put them back to the final reduced system.

We now describe how the decomposition can be computed using the matrix sign function. The matrix sign function of Z is defined as

$$sign(Z) := S \begin{bmatrix} -I_k & 0 \\ 0 & I_{n-k} \end{bmatrix} S^{-1}. \tag{3.60}$$

If an eigenvalue $\lambda > 0$, then its corresponding eigenvalue in $\mathrm{sign}(Z)$ is $+1$ and if an eigenvalue $\lambda < 0$, then its corresponding eigenvalue in $\mathrm{sign}(Z)$ is -1. Note that $\mathrm{sign}(Z)$ is unique and independent of the order of the eigenvalues in the Jordan decomposition of Z.

Consider the realization (A, B, C, D) of a continuous-time linear time-invariant system, and let $sign(A)$ denote the sign function of A. We start by computing a QR factorization of $I_n - sign(A)$ as

$$\begin{aligned} I_n - sign(A) &= QRP, \\ R &= \begin{bmatrix} R_{11} & R_{12} \\ 0 & 0 \end{bmatrix}, \end{aligned} \tag{3.61}$$

where $Q \in R^{n\times n}$ is orthogonal, $R \in R^{n\times n}$ is upper triangular, with $R_{11} \in R^{k\times k}$, and $P \in R^{n\times n}$ is a permutation matrix. Note that all the positive eigenvalues of A will become 2 in $I_n - sign(A)$, while all the negative eigenvalues of A will become zero theoretically as shown in the last $n-k$ rows of R. But they may not be exactly zero numerically. So the zeros in the last $n-k$ rows of R are to be understood as "zero with respect to a given tolerance threshold." Then the first k columns of Q span the stable A-invariant subspace. Thus,

$$\tilde{A} := Q^T A Q = \begin{bmatrix} A_{11} & A_{12} \\ 0 & A_{22} \end{bmatrix}, \tag{3.62}$$

where $\Lambda(A_{11}) = \Lambda(A) \cap C^-$, and $\Lambda(A_{22}) = \Lambda(A) \cap C^+$, $\Lambda(X)$ means the eigenvalues of X. In a second step, we compute a matrix $V \in R^{n\times n}$ such that

$$\hat{A} := V^{-1}\tilde{A}V = \begin{bmatrix} I_k & -Y \\ 0 & I_{n-k} \end{bmatrix} \begin{bmatrix} A_{11} & 0 \\ 0 & A_{22} \end{bmatrix} \begin{bmatrix} I_k & Y \\ 0 & I_{n-k} \end{bmatrix}, \tag{3.63}$$

where $Y \in R^{k\times n-k}$ satisfies the Sylvester equation

$$A_{11}Y - YA_{22} + A_{12} = 0. \tag{3.64}$$

As $\Lambda(A_{11}) \cap \Lambda(A_{22}) = \emptyset$, the equation has a unique solution. Sylvester equations with strictly stable or unstable coefficient matrices can be solved using the iterative algorithm [10].

The desired additive decomposition of $G(s) = C(sI - A)^{-1}B + D$ is finally obtained by performing the state-space transformation

$$\begin{aligned}(\hat{A}, \hat{B}, \hat{C}, \hat{D}) :&= (V^{-1}Q^TAQV, V^{-1}Q^TB, CQV, D) \\ &= \left(\begin{bmatrix} A_{11} & 0 \\ 0 & A_{22} \end{bmatrix}, \begin{bmatrix} B_1 \\ B_2 \end{bmatrix}, [C_1\ C_2], D \right),\end{aligned} \tag{3.65}$$

where $\hat{A}, \hat{B}, \hat{C}$ are partitioned conformally so that

$$\begin{aligned} G(s) &= C(sI - A)^{-1}B + D \\ &= \hat{C}(sI - \hat{A})^{-1}\hat{B} + \hat{D} \\ &= C_1(sI_k - A_{11})^{-1}B_1 + D + C_2(sI_{n-k} - A_{22})^{-1}B_2 \\ &= G_-(s) + G_+(s), \end{aligned} \tag{3.66}$$

where $G_-(s)$ is a stable part and $G_+(s)$ is a unstable part of transfer function matrix. Then we can proceed to perform the TBR on the $G_-(s)$.

3.9.2 *Generalized eigenproblem method for AREs*

When solving the ARE equation in (3.45) for linear VLSI interconnect RLC circuits, the directly coupling part D matrix is typically zero (or singular). As a result, we cannot solve (3.45) in case D matrix is zero or singular. This section presents numerical methods to solve AREs for both non-singular and singular D cases.

Solutions of the Lur'e equations and solutions of algebraic Riccati equations are closely related. We summarize the basic numerically robust computational procedure below [103].

ARE solution for non-singular $D + D^T$

Let F, X, Y be real $n \times n$ matrices with X and Y symmetric and positive semidefinite. Assume that there exists a symmetric matrix $P \in R^{n\times n}$ that satisfies the Riccati equation

$$F^TP + FP + PXP + Y = 0, \tag{3.67}$$

and that all the eigenvalues of $F + XP$ are in C^- (i.e., they have (strictly) negative real parts). Then it follows that such a P is unique and we shall use the notation $Ric(F, X, Y)$ to refer to it. $P = Ric(F, X, Y)$ can be determined by finding the eigenvectors of the matrix

$$M := \begin{bmatrix} F & X \\ -Y & -F^T \end{bmatrix} \in R^{2n\times 2n}, \tag{3.68}$$

which has n eigenvalues in C^- and n in C^+. Let the eigenvectors corresponding to the eigenvalues in C^- be denoted by $w_1, \cdots, w_n$, and define $W := [w_1, \cdots, w_n] \in R^{2n\times n}$. Partitioning W as $W = \begin{bmatrix} W_1 \\ W_2 \end{bmatrix}$, then it follows that $W_1 \in R^{n\times n}$ is non-singular and $Ric(F, X, Y) = W_2 W_1^{-1}$. Following is the algorithm for the computation of P matrix for a positive-real system (A, B, C, D) with non-singular $D + D^T$.

Algorithm 3.5: COMPUTATION OF P WITH NON-SINGULAR $D + D^T$

1. Let $F = A - BKC$, $K = (D + D^T)^{-1}$.
2. Determine $P = Ric(F, BKB^T, C^T KC)$.

ARE solution for singular $D + D^T$

Now we consider constructing a positive real (PR) solution algorithm for the case of singular $D + D^T$. In Algorithm 3.6, the system (A, B, C, D) is first transformed to a system (A_I, B_I, C_I, D_I) with non-singular $D_I + D_I^T$ to which Algorithm 3.5 is applicable. The P matrix of the original system is then computed based upon those of (A_I, B_I, C_I, D_I) [103]. Following is the algorithm for the computation of P matrix in the positive-real lemma for an positive-real system (A, B, C, D) with $r := \text{rank}(D + D^T) < m$

Algorithm 3.6: COMPUTATION OF P MATRIX WITH SINGULAR $D + D^T$

1. Perform the singular value decomposition of $\bar{D} := (D + D^T)/2$:

$$\bar{D} = U \begin{bmatrix} \Sigma & 0 \\ 0 & 0 \end{bmatrix} U^T, \tag{3.69}$$

and set

$$R := U \begin{bmatrix} \Sigma^{-\frac{1}{2}} & 0 \\ 0 & I_{(m-r)\times(m-r)} \end{bmatrix}. \tag{3.70}$$

Then transform $\bar{D}$ to

$$\hat{D} := R^T \bar{D} R = \begin{bmatrix} I_{r\times r} & 0 \\ 0 & 0 \end{bmatrix}. \tag{3.71}$$

2. Partition $B_R := BR$ and $C_R := R^T C$ into

$$B_R = \begin{bmatrix} B_1 \ B_2 \end{bmatrix}, \tag{3.72}$$

$$C_R = \begin{bmatrix} C_1 \\ C_2 \end{bmatrix}, \tag{3.73}$$

with $B_1 \in R^{r\times n}$ and $C_1 \in R^{r\times n}$. Pick $n - m + r$ linearly independent vectors $v_1, \cdots, v_{n-m+r}$ that span the null space of C_2. Set $V := [v_1, \cdots, v_{n-m+r}]$ and $T := [B_2 V]$. Note that T is always invertible since $C_2 B_2$ is invertible. Partition T^{-1} into

$$T^{-1} = \begin{bmatrix} B_2^{\#} \\ V^{\#} \end{bmatrix}, \tag{3.74}$$

$$B_2^{\#} = (C_2 B_2)^{-1} C_2. \tag{3.75}$$

3. Transform $(A, B, C, \bar{D})$ to $(\hat{A}, \hat{B}, \hat{C}, \hat{D})$ by the positive-real preserving transformation (T, R):

$$\hat{A} := T^{-1} A T = \begin{bmatrix} \hat{A}_{11} \ \hat{A}_{12} \\ \hat{A}_{21} \ \hat{A}_{22} \end{bmatrix}, \tag{3.76}$$

$$\hat{B} := T^{-1} B_R = \begin{bmatrix} \hat{B}_{11} \ I_{(m-r)\times(m-r)} \\ \hat{B}_{21} \quad 0 \end{bmatrix}, \tag{3.77}$$

$$\hat{C} := C_R T = \begin{bmatrix} \hat{C}_{11} \ C_{12} \\ \hat{C}_{21} \ \ 0 \end{bmatrix}, \tag{3.78}$$

$$\hat{D} := \begin{bmatrix} \hat{I}_{r\times r} \ 0 \\ 0 \ \ 0 \end{bmatrix}. \tag{3.79}$$

Go to the next step only if $\hat{C}_{21} = C_2 B_2$ is symmetric and positive definite; otherwise the system in question is not positive real.

4. Determine the state-space realization (A_I, B_I, C_I, D_I) of the partial inverse of $(\hat{A}, \hat{B}, \hat{C}, \hat{D})$ as follows:

$$A_I := \hat{A}_{22} \in R^{(n-m+r)\times(n-m+r)}, \tag{3.80}$$

$$B_I := [\hat{B}_{21} \hat{A}_{21}], \tag{3.81}$$

$$C_I := \begin{bmatrix} \hat{C}_{12} \\ -\hat{C}_{21} \hat{A}_{12} \end{bmatrix}, \tag{3.82}$$

$$D_I := \begin{bmatrix} I_{r\times r} & C_1 B_2 \\ -C_2 B_1 & -C_2 A B_2 \end{bmatrix}. \tag{3.83}$$

Go to the next step only if D_I is positive definite; otherwise the system in question is not positive real.

5. Use Algorithm 3.5 to find the positive-real matrix, P_I, for (A_I, B_I, C_I, D_I) where $D_I + D_I^T$ is non-singular:

n	σ	n	σ	n	σ
1	∞	11	0.00074038	21	1.8735×10^{-9}
2	32.643	12	0.00040137	22	7.025×10^{-10}
3	6.7536	13	7.6789×10^{-5}	50	5.2306×10^{-19}
4	2.0489	14	3.0178×10^{-5}	100	3.1665×10^{-20}
5	0.71765	15	1.0206×10^{-5}	150	3.157×10^{-21}
6	0.40582	16	2.323×10^{-6}	200	3.8209×10^{-23}
7	0.10189	17	8.8947×10^{-7}	250	2.5916×10^{-27}
8	0.023015	18	1.1729×10^{-7}	300	8.5601×10^{-36}
9	0.009922	19	4.2241×10^{-8}	400	1.1436×10^{-52}
10	0.0031625	20	1.1495×10^{-8}	516	3.2594×10^{-89}

Table 3.1 *The Hankel singular values for a six-port linear interconnect circuit.*

$$P_I = Ric(F_I, B_I K_I B_I^T, C_I^T K_I C_I), \tag{3.84}$$

where $K_I := (D_I + D_I^T)^{-1}$ and $F_I := A_I - B_I K_I C_I$.

6. Compute P according to

$$P := V^{\#^T} P_I V^{\#} + B_2^{\#^T} \hat{C}_{21} B_2^{\#}. \tag{3.85}$$

If matrix $D = 0$, then all the submatrices in the above algorithm with r number of rows or columns are removed from their main matrix, e.g., $B_R = [B_1 B_2]$, $B_1 \in R^{n \times r}$ becomes $B_R = B = B_2$ if $r = 0$.

3.10 Numerical examples

In this section, we present some numerical results of TBR-based reduction of an interconnect circuit with six ports from a practical industry chip.

The RLC interconnect circuit has 516 states in which there exist one unstable state (pole). So its stable part is isolated, balanced. The entries corresponding to unstable modes are set to ∞.

After balancing the system, we list the Hankel singular values in Table 3.1 in descending order where n indicates the singular value indices and *sigma* denotes the singular value.

From Table 3.1, we can observe that the Hankel singular values of the system decay extremely rapidly. As a matter of fact, many RLC interconnect circuits have such a property. Hence, very low rank approximations are possible and accurate low-order reduced models can be obtained.

After the truncation, the resulting Gramians contain the largest k singular values σ_1 through σ_k ($k = 1, 2, 3, 4, 5$, respectively).

The approximation results with different singular values (states) are shown in

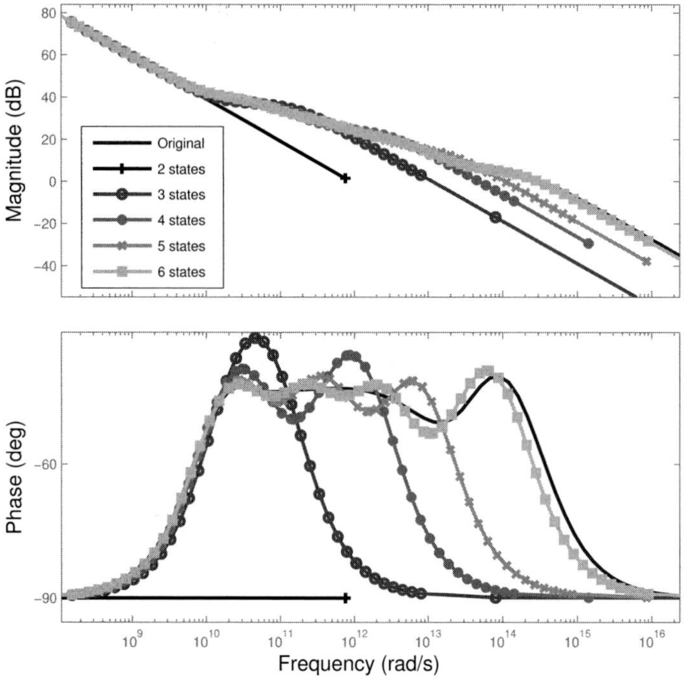

Figure 3.1 Frequency responses of a reduced model and its original system.

Fig. 3.1. From Fig. 3.1, we observe that first the more states we retain, the better both curves can match. Second, even a reduced model, which retains a small number of dominant states (six in this case), can be accurate to approximate the original system with a large number of states (516 in this case).

Next, we use the ten largest states to approximate the original system. The frequency response of the input impedance for port 1 is shown in Fig. 3.2. We can see that the two responses are almost the same up to 10^{16} Hz.

3.11 Summary

In this chapter, we have introduced truncated balanced realization methods for model order reduction of general linear time-invariant dynamic systems. TBR method is a the singular-value-decomposition-based approaches. In the TBR methods, the system is first transformed into an eigenspace of the balanced controllability and observability Gramian. Then SVD is performed and states with weak controllability and observability are dropped. We also discussed positive real TBR methods, which can produce passive models for passive circuits but at the expense of high computation costs. To avoid the high computation costs associated with solving

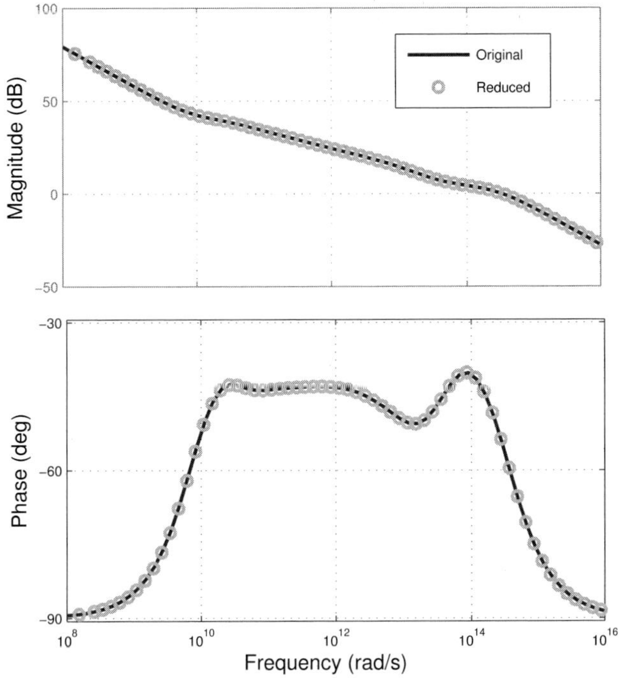

Figure 3.2 Frequency response of the input impedance of a reduced model and its original system.

Lyapunov equations for classic TBR methods, an empirical TBR method was also presented, where Gramians are computed by sampling the frequency-domain state impulse responses. Finally we discuss some numerical issues associated with implementations of TBR methods and see how those issues are resolved in the solving of Lyapunov equations and algebraic Riccati equations.

4 Passive balanced truncation of linear systems in descriptor form

4.1 Introduction

In Chapter 3, we introduced the truncated balanced realization (TBR) method for compact modeling of interconnect circuits. In this chapter, we introduce a new passive TBR method, which only requires the solving of generalized Lyapunov equations and is able to preserve the structure of the reduced models just like the structure-preserving Krylov subspace method of [37].

In the TBR method, two steps are involved in the reduction process: the *balancing* step aligns the states such that states can be controlled and observed equally. The *truncating* step then throws away the weak states, which usually leads to much smaller model. The major advantage of TBR methods is that TBR methods can give deterministic global bound for the approximate error and can give nearly optimal models in terms of errors and model sizes [60].

Standard TBR algorithms, which solve the Lyapunov equations (linear matrix equations) do not necessarily preserve passivity. To ensure passivity, positive-real TBR (PR-TBR) has to be carried out [88, 131] by solving more difficult Lur'e or Riccati equations, which can be computationally prohibitive as they are non-linear (quadratic) matrix equations.

Given a state-space model in *descriptor* form

$$\begin{aligned} E\tfrac{\mathrm{d}x}{\mathrm{d}t} &= Ax(t) + Bu(t), \\ y(t) &= Cx(t) + Du(t), \end{aligned} \tag{4.1}$$

where $E,\ A \in R^{n\times n}$, $B \in R^{n\times p}$, $C \in R^{p\times n}$, $D \in R^{p\times p}$, and $y(t),\ u(t) \in R^p$. When $E = I$, (4.1) is in standard state-space form. Note that the descriptor form is the natural form of the circuit MNA matrices for interconnect circuits, where E is the matrix of storage elements, A is the matrix of conductance and $B = C^T$ is the input and output position matrices and $D = 0$.

The existing passive TBR methods first convert original descriptor systems into standard state-space equations by mapping $E \rightarrow I$, $A \rightarrow E^{-1}A$, and $B \rightarrow E^{-1}B$ and then solving two Lur'e or Riccati equations to guarantee the passivity of reduced model. However, there are several issues related to PR-TBR: first, it is not numerically reliable in the sense that given an ill-conditioned E, the mapping can generate too much numerical error and sometimes even the stability of the system cannot be guaranteed in this process. Second, Lur'e and Riccati equations are quadratic matrix equations and thus usually more expensive than Lyapunov equations, which are linear matrix equations. Third, the structure property (symmetry, sparsity) inherent to RLC circuits cannot be preserved after the reduction. Fourth, in most cases, PR-TBR is not as accurate as standard TBR [88].

In this chapter, we present a novel passivity-preserving TBR method, named *PriTBR*, for interconnect modeling. Instead of working on the standard state-space equations, PriTBR works on space equations in descriptor form directly by solving generalized Lyapunov equations. Since the circuit matrix structure is preserved in the state-space equations in the descriptor form, congruence transformation can be applied to ensure the passivity of reduced models. Compared with existing PR-TBR, which solves more difficult Lur'e or Riccati equations on standard state-space equations, the new method is numerically reliable, less expensive, and more accurate. The new method has the same accuracy as the standard TBR method for RC circuits using the first-order formulation and can easily be extended to RLCM circuits to preserve the structure information of RLC circuit matrices using the second-order formulation. By combining it with structure-preserving Krylov-subspace based MOR methods, it can generate optimal structure preserved reduced model from large-scale circuits.

4.2 The passive balanced truncation algorithm: PriTBR

The new approach is motivated by recent balanced truncation work on the linear systems in descriptor form [119] as shown in (4.1). But the existing method does not give passive reduction.

In the PriTBR method, we work directly on systems in descriptor form, which is the natural form of RLC circuits in MNA formulation. Instead of obtaining a balanced form and truncating, we compute the basis, which spans the dominant subspace corresponding to the first q largest Hankel singular values. After that, we orthonormalize it and project the system onto it so that the reduction process can be viewed as a congruence transformation. The difference between the new method and the standard TBR method is just like the difference between PRIMA [85] and Pade approximation via Lanczos (PVL) [32].

4.2.1 *Generalized balanced truncation by projection*

Given a state-space model in descriptor form in (4.1) with the stable matrix pencil $\lambda E - A$, which is usually the case in an RLC circuit, we first assume that E is non-singular. This restriction can easily be released with some additional steps [119]. If the system is in descriptor form, the controllable and observable Gramians can be computed by solving generalized Lyapunov equations [119].

$$\begin{aligned} EPA^T + APE^T &= -BB^T, \\ E^TQA + A^TQE &= -C^TC. \end{aligned} \tag{4.2}$$

The matrix PE^TQE has non-negative eigenvalues, and the square roots of these eigenvalues $\sigma_j = \sqrt{\lambda_j(PE^TQE)}$ define the Hankel singular values of the system. We assume that σ_j are ordered decreasingly. The system is called balanced if

$$P = Q = diag(\sigma_1, \sigma_2, \ldots, \sigma_n) \tag{4.3}$$

PE^TQE here is similar to W_cW_o in the standard TBR method. After solving the generalized Lyapunov equations, one can compute the similarity transformation matrix T in (3.26) based on the square-root method [68]. The subsequent steps in [119] are the same as the TBR method in standard form: to first balance the system and then truncate it.

However, in this paper, instead of obtaining a balanced form and truncating, we perform reduction from the viewpoint of projection. As we know in standard TBR, the similarity transformation matrix T is the eigenmatrix of PE^TQE

$$TPE^TQET^{-1} = diag({\sigma_1}^2, {\sigma_2}^2, \ldots, {\sigma_n}^2) \tag{4.4}$$

Then, the balancing and truncation can be viewed as a special projection of the system onto the dominant eigenspace of the matrix PE^TQE corresponding to the q largest eigenvalues. Those eigenvalues are more important in terms of input–output behavior of the system. The dominant subspace is spanned by the first q columns of matrix T. If we partition T into

$$T = \begin{bmatrix} T_1 \; T_2 \end{bmatrix}, \tag{4.5}$$

where $T_1 \in R^{n\times q}$, $T_2 \in R^{n\times(n-q)}$. Then the dominant subspace is spanned by basis T_1. Orthonormalize the basis

$$X = \text{orth}(T_1), \tag{4.6}$$

where $X \in R^{n\times q}$. $X^TX = I$ and X spans the same subspace as T_1

$$\text{span}(X) = \text{span}(T_1). \tag{4.7}$$

After that, we perform the reduction directly by projecting the system onto the subspace spanned by X

$$x = X\tilde{x} \tag{4.8}$$

where $\tilde{x} \in R^q$ is the reduced system variable. The reduced-order system matrices are

$$\tilde{A} = X^TAX,\ \tilde{E} = X^TEX,\ \tilde{B} = X^TB,\ \tilde{C} = CX. \tag{4.9}$$

These types of transformation are known as congruence transformations. In the following, we review congruence transformation and its passivity-preserving property.

4.2.2 *Passive reduction through congruence transformation*

For RLC interconnect circuits, we can formulate the original circuit matrix with MNA formulation into the so-called passive form [85]:

$$\begin{aligned} C\tfrac{\mathrm{d}x}{\mathrm{d}t} &= -Gx(t) + Bu(t), \\ y(t) &= B^T x(t), \end{aligned} \tag{4.10}$$

such that the conductance matrix and the storage element matrix are positive semidefinite (.i.e., $G \geq 0$ and $C \geq 0$) and the input and output position matrices are the same. Notice that such passive form is also the descriptor form in (4.1). Then the transfer function of the model will be positive real, meaning that the model is passive.

$$\tilde{C} = X^T C X, \ \tilde{G} = X^T G X, \ \tilde{B} = X^T B. \tag{4.11}$$

Since congruence transformation preserves the definiteness of the matrix, the reduced $\tilde{G}$, $\tilde{C}$ are still positive semidefinite. Then the transfer function of the reduced model will be positive real, and thus passive. Therefore, PriTBR can preserve the passivity for general RLC circuits.

For large systems, direct application of balanced truncation is computationally infeasible. Therefore, the methods are of more interest when combined with iterative Krylov-subspace procedures like PRIMA. Since an initial reduced model via PRIMA still has this convenient passive form, PriTBR can always be used as the second stage of a composite model reduction procedure to generate a passive preserved reduced model.

4.2.3 *Comparison with PR-TBR*

Compared with passive PR-TBR method, this process has the following advantages:

First, we do not need to put a descriptor system into a standard form by mapping $E \rightarrow I$, $A \rightarrow E^{-1}A$, $B \rightarrow E^{-1}B$. After an initial projection MOR step (assuming Krylov-subspace method is performed as a first stage), E is usually non-singular [88] but maybe ill-conditioned. As a result, the result by PR-TBR is no more accurate from the numerical point of view. Even worse, sometimes, an unstable system can be generated after the mapping, so that the Lur'e equations do not have positive semidefinite solutions.

Second, the Lur'e equation is a quadratic matrix equation, which is more expensive than linear matrix equations like the Lyapunov equation. The generalized Lyapunov equation is still a linear matrix equation; when E is non-singular, it costs almost the same as the Lyapunov equation [119].

Third, PriTBR has the same accuracy as standard TBR but PR-TBR is usually not so accurate as standard TBR [88].

Fourth, as shown in the next section, PriTBR can be generalized to preserve structure and reciprocity.

However, like PRIMA, it cannot be used on systems of non-RLC circuits because both of them rely on congruence transformations to preserve passivity.

4.2.4 PriTBR reduction algorithm

In this section, we present the complete PriTBR reduction flow. On top of PriTBR, we also present the combined PriTBR and PRIMA flow.

Algorithm 4.1: PROJECTION-BASED PASSIVE TBR (PRITBR)

1. Solve $EPA^T + APE^T = -BB^T$ for P.
2. Solve $E^TQA + A^TQE = -C^TC$ for Q.
3. Compute Cholesky factors $P = L_pL_p^T,\ Q = L_QL_Q^T$.
4. Compute SVD of $L_p^TE^TL_Q$:

$$L_p^TE^TL_Q = [U_1, U_2]\begin{bmatrix}\Sigma_1 & 0\\ 0 & \Sigma_2\end{bmatrix}[V_1, V_2]^T. \tag{4.12}$$

5. Compute the dominant basis $T = L_pU_1\Sigma_1^{-1/2}$.
6. Orthonormalize T: $X = \text{orth}(T)$.
7. Compute the reduced system with $\tilde{E} = X^TEX$; $\tilde{A} = X^TAX$; $\tilde{B} = X^TB$; $\tilde{C} = CX$.

The basic algorithm is a generalization of the square-root method used in standard balanced truncation [68].

The new method can also be combined with a projection-based MOR method like PRIMA to deal with large scale circuits. This is similar to the method in [60].

Algorithm 4.2: COMBINED PRIMA AND PRITBR MODEL ORDER REDUCTION

1. Perform PRIMA to obtain a small-size passive-preserved initial reduced model from a large-size original model.
2. Perform PriTBR to obtain an optimal passive-preserved final reduced model from the initial model.

4.3 Structure-preserved balanced truncation

While PR-TBR generates provably passive reduced-order models, it does not preserve other structures, such as reciprocity (symmetry) or the block structures of the circuit matrices, inherent to RLC circuits. However, the PriTBR in the previous section can easily be extended such that both passivity and structure can be preserved in a balanced truncation process.

4.3.1 Structure-preserved balanced truncation

Similar to the SPRIM [37], we assume that only current sources are applied and the transfer function is an impedance matrix function.

$$Z(s) = B^T(G + sC)^{-1}B. \tag{4.13}$$

In the modified nodal formulation (MNA) of an RLC circuit, G, C, B have the block structure

$$G = \begin{bmatrix} G_1 & G_2^T \\ -G_2 & 0 \end{bmatrix}, C = \begin{bmatrix} C_1 & 0 \\ 0 & C_2 \end{bmatrix}, B = \begin{bmatrix} B_1 \\ 0 \end{bmatrix}. \tag{4.14}$$

Let V be the matrix T_1 used in PriTBR. Let

$$V = \begin{bmatrix} V_1 \\ V_2 \end{bmatrix} \tag{4.15}$$

be the partitioning of V corresponding to the block sizes of G and C. We then split the projection matrix V and orthonormalize each block respectively as

$$\tilde{V} = \begin{bmatrix} \mathrm{orth}(V_1) & 0 \\ 0 & \mathrm{orth}(V_2) \end{bmatrix}. \tag{4.16}$$

Notice that

$$\mathrm{span}(V) \subseteq \mathrm{span}(\tilde{V}). \tag{4.17}$$

We, therefore, can project onto $\tilde{V}$, which spans at least the same dominant subspace as V. As shown in experiments, it has the good properties of the standard balanced truncation and matches the original model well globally.

At first glance, SP-PriTBR is not optimal in model size compared with PriTBR because, given the same error bound, the SP-PriTBR model would be twice as large as the corresponding PriTBR model. However, the SP-PriTBR model can always be represented in the second-order form. In this sense, the SP-PriTBR model (when written in second-order form) has the same dimension as PriTBR model given the same error bound,

$$\begin{aligned} &\tilde{G}_1 = V_1^T G_1 V_1, \tilde{G}_2 = V_2^T G_2 V_1, \tilde{C}_1 = V_1^T C_1 V_1, \\ &\tilde{C}_2 = V_2^T C_2 V_2, \tilde{B}_1 = V_1^T B_1 \end{aligned} \tag{4.18}$$

and

$$\tilde{G} = \begin{bmatrix} \tilde{G}_1 & \tilde{G}_2^T \\ -\tilde{G}_2 & 0 \end{bmatrix}, \tilde{C} = \begin{bmatrix} \tilde{C}_1 & 0 \\ 0 & \tilde{C}_2 \end{bmatrix}, \tilde{B} = \begin{bmatrix} \tilde{B}_1 \\ 0 \end{bmatrix}. \tag{4.19}$$

The reduced model $\tilde{Z}_n$ in first-order form,

$$\tilde{Z}(s) = \tilde{B}^T (\tilde{G} + s\tilde{C})^{-1} \tilde{B}, \tag{4.20}$$

and in second-order form,

$$\tilde{Z}(s) = \tilde{B}_1^T (s\tilde{C}_1 + \tilde{G}_1 + \frac{1}{s}\tilde{G}_2^T \tilde{C}_2^{-1} \tilde{G}_2)^{-1} \tilde{B}_1, \tag{4.21}$$

and

$$\tilde{C}_1 \succeq 0, \tilde{G}_1 \succeq 0, \tilde{G}_2^T \tilde{C}_2^{-1} \tilde{G}_2 \succeq 0, \tag{4.22}$$

and thus $\tilde{Z}(s)$ is passive and also symmetric. This means that the reduced model preserves reciprocity and can be more easily synthesized as an actual circuit.

4.3.2 Structure-preserving PriTBR algorithm

Algorithm 4.3: STRUCTURE-PRESERVED PRITBR (SP-PRITBR)

1. Perform Algorithm 4.2 (step 1 – step 5) for $V = T_1$.
2. Partition V corresponding to the block sizes of G and C,

$$V = \begin{bmatrix} V_1 \\ V_2 \end{bmatrix}. \tag{4.23}$$

3. Set

$$\tilde{V} = \begin{bmatrix} \mathrm{orth}(V_1) & 0 \\ 0 & \mathrm{orth}(V_2) \end{bmatrix}. \tag{4.24}$$

4. Obtain the reduced model by projection:

$$\tilde{G} = \tilde{V}^T G \tilde{V}, \tilde{C} = \tilde{V}^T C \tilde{V}, \tilde{B} = \tilde{V}^T B. \tag{4.25}$$

The SP-PriTBR algorithm can also work together with the structure-preserving projection-based method like SPRIM to produce a structure-preserved compact model from large scale circuits as shown in the following algorithm:

Algorithm 4.4: COMBINED SPRIM AND SP-PRITBR MODEL ORDER REDUCTION

1. Perform SPRIM to obtain a small-size structure-preserved initial reduced model from a large-size original model.
2. Perform SP-PriTBR to get an optimal structure-preserved final reduced model from the first step.

4.4 Numerical examples

In this section, we show examples that illustrate the effectiveness of proposed PriTBR methods and compare them with existing relevant approaches.

4.4.1 The accuracy of PriTBR

First, we demonstrate empirically that PriTBR has the same accuracy as standard balanced truncation for RC circuits. Compared with PRIMA, both of them are optimal in the sense that given a reduced order, they are more accurate. Here, the original model is a 515-order RC circuit stimulated by a current source. In Figure 4.1, given the same reduced order ten, both PriTBR and standard balanced truncation match equally well with the original curve and far better than PRIMA of the same reduced order.

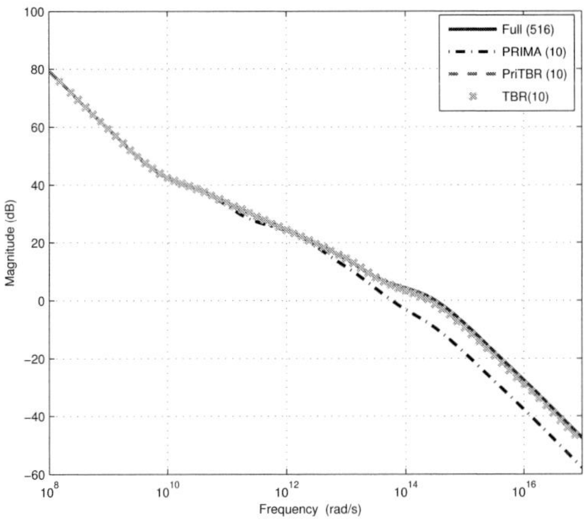

Figure 4.1 Frequency responses of TBR, PriTBR, and PRIMA reduced models and the original circuit.

4.4.2 *The guaranteed passivity of PriTBR*

The second example is an RLC transmission line (order 904) with voltage source input. The initial reduction is done by PRIMA and then followed by a TBR and PriTBR, respectively. The final reduced order is 12.

In Figure 4.2, the Nyquist plot of the driving-point impedance is shown. This Nyquist plot contains both the magnitude and phase information about the network impedance. It also provides a graphical test of port passivity. Indeed, it is well known that the Nyquist plots of positive-real transfer functions lie entirely in the right half of the complex plane.

At first glance, it seems that the reduced models are both passive. But when zooming in, we find that, for PriTBR, the entire Nyquist plot lies in the right half of the complex plane. However, for standard balanced truncation, the Nyquist plot extends to the left half of the complex plane, which means that passivity is not preserved in the reduced model.

4.4.3 *The numerical stability and accuracy of reduced model by PriTBR*

We use an RC circuit of 517 order with a voltage source input to demonstrate the advantage of PriTBR. We do a initial reduction by PRIMA and obtain a reduced model of 75 order, which matches the original curve very well.

First, if we want to use PR-TBR for the second stage, we need map $E \to I$, $A \to E^{-1}A$, $B \to E^{-1}B$ to put a descriptor system into a standard form. We find that after the mapping, the standard form is no more stable and thus no positive semidefinite solution is available for the following Lur'e equations. Figure 4.3 is the pole-zero map of the system around the origin before and after mapping. We can

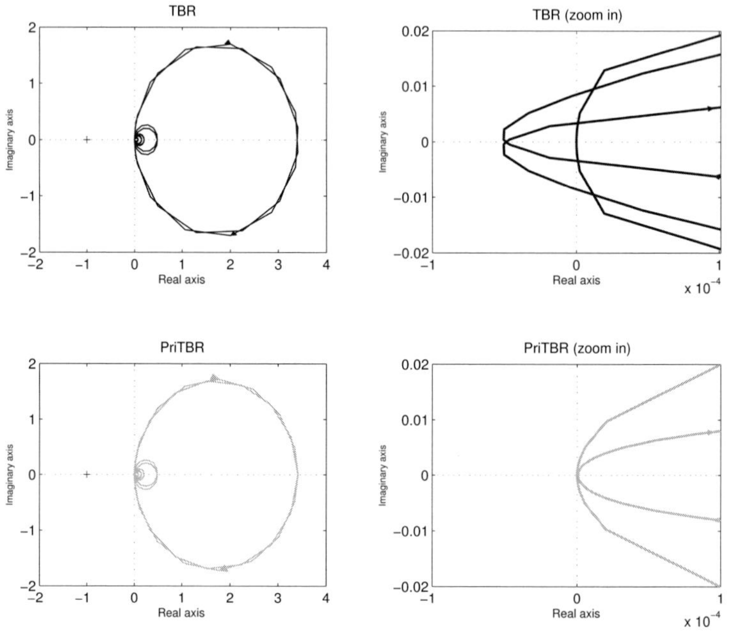

Figure 4.2 Nyquist plots of the TBR reduced model and the PriTBR reduced model.

see after mapping a positive pole is generated.

Then, we employ PriTBR as the second reduction stage and get a final reduced model of 15 order, which is still indistinguishable with the original curve. However, if we use PRIMA to get a reduced model of 15 directly, we find that the difference is very obvious.

4.4.4 Comparison of SPRIM and SP-PriTBR

We use an RLC circuit of 302 order with current source input to compare SPRIM and SP-PriTBR with the same reduced order ten. The structure inherent to RLC circuit is preserved in both of them. We find that SP-PriTBR inherits the optimal property of standard balanced truncation and matches the original curve better than SPRIM in a wide frequency band.

4.5 Summary

In this chapter, we presented a novel passive projection-based balanced truncation model reduction method, named *PriTBR*. The new method combines the TBR method with a projection framework to produce passive models for the first time.

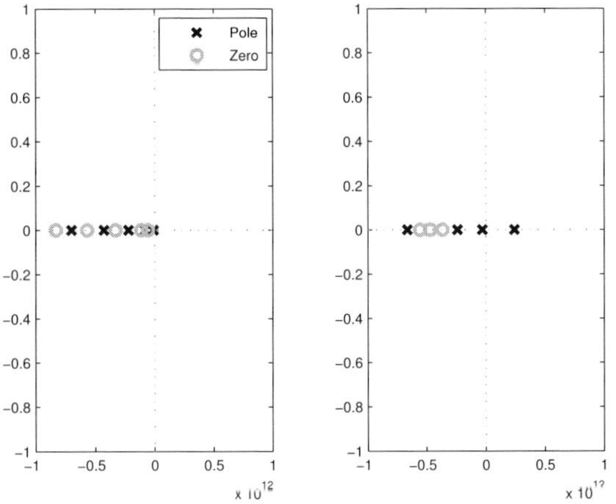

Figure 4.3 Pole zero map of system before mapping.

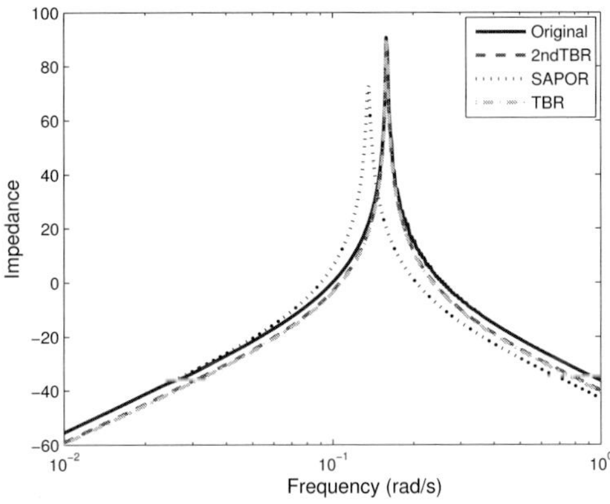

Figure 4.4 Frequency responses of PRIMA and combined PRIMA and PriTBR reduced models and the original circuit.

It enjoys the good error bounds from the TBR method and passive-reduction benefit from congruence transformation. Compared with existing passive TBR, the new technique is numerically reliable, more accurate and less expensive. In addition to passivity-preservation, we also showed that PriTBR can be extended to preserve structure information such as reciprocity and block structure. It can be applied as a second stage of model order reduction to work with Krylov subspace methods to generate a nearly optimal reduced model from a large-scale interconnect circuit while passivity, structure, and reciprocity can be preserved at the same time. Experimental results demonstrated the effectiveness of the new method and its advantage over existing passive TBR and projection-based MOR methods.

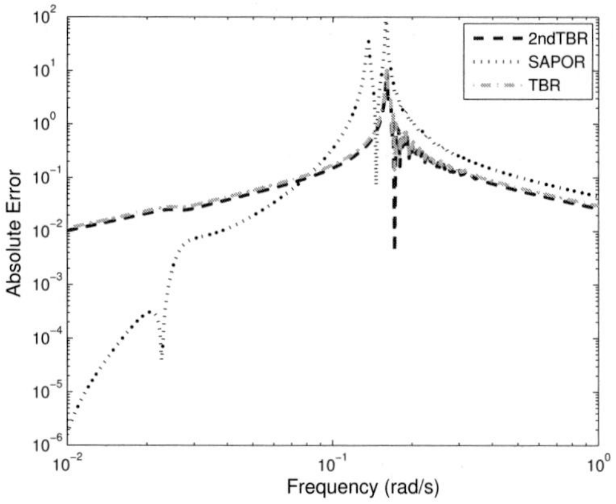

Figure 4.5 Frequency responses of SPRIM and SP-PriTBR reduced models and the original circuit.

5 Passive hierarchical model order reduction

In this chapter, we focus on passive wideband modeling of RLCM circuits. We propose a new passive wideband reduction and realization framework for general passive high-order RLCM circuits. Our method starts with large RLCM circuits, which are extracted by existing geometry extraction tools like FastCap [83] and FastHenry [59] under some relaxation conditions of the full-wave Maxwell equations (like electro-quasi-static for FastCap or magneto-quasi-static for FastHenry) instead of measured or simulated data. It is our ultimate goal that we can obtain the compact models directly from complex interconnect geometry without measurement or full-wave simulations. The method presented in this chapter is called hierarchical model order reduction, *HMOR*, which is based on the general frequency-domain hierarchical model reduction algorithm [121,122,124] and an improved VPEC (vector potential equivalent circuit) [134] model for self and mutual inductance, which can be easily sparsified and is hierarchical-reduction friendly.

The HMOR method achieves passive wideband modeling of RLC circuits via multi-point expension and the convex programming based passivity enforcement method. In this section, we will show that the frequency-domain hierarchical reduction is equivalent to implicit moment-matching around $s = 0$, and that the existing hierarchical reduction method by one-point expansion [121, 124] is numerically stable for general tree-structured circuits. We also show that HMOR preserves reciprocity of passive circuit matrices. We present a hierarchical multi-point reduction scheme to obtain accurate order-reduced admittance matrices of general passive circuits. An explicit waveform matching algorithm is applied for searching dominant poles and residues from different expansion points based on the unique hierarchical reduction framework.

To enforce passivity, the state-space-based convex programming optimization technique [21] is applied to the model order reduced admittance matrix, which is to be presented in Chapter 10. To realize the passivity-enforced admittance, we use a general, reciprocity-preserving, passivity-preserving, multi-port network realization method based on a relaxed one-port network synthesis technique using Foster's canonical form in an error-free manner, which is to be represented in Chapter 11. The resulting modeling algorithm can take in general RLCM SPICE netlists and generate SPICE netlists of passive multi-port models for any linear passive network with easily controlled model accuracy and complexity.

5.1 Overview of hierarchical MOR algorithm

In this section, we first briefly review hierarchical subcircuit reduction and the one-point frequency-domain hierarchical model order reduction method [122] before we present the multi-point hierarchical model order reduction method, *HMOR*. Then we give the flow of the hierarchical model order reduction and realization method.

5.1.1 Review of hierarchical subcircuit reduction

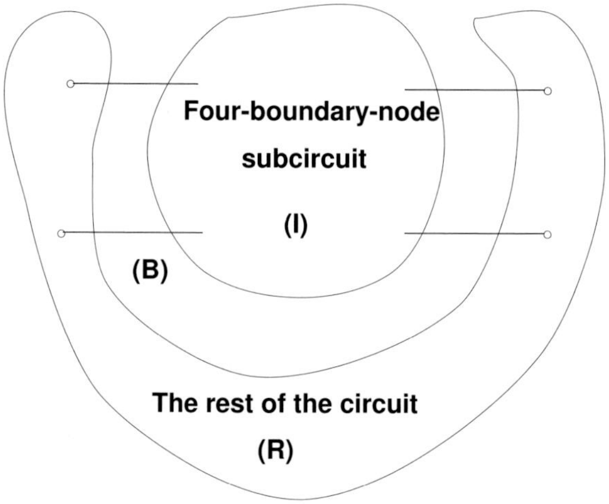

Figure 5.1 A hierarchical circuit. Reprinted with permission from [126] (c) 2000 IEEE.

Consider a subcircuit with some internal structures and terminals, as illustrated in Figure 5.1. The circuit unknowns — the node-voltage variables and branch-current variables — can be partitioned into three disjoint groups x^I, x^B, and x^R, where the superscripts *I, B, R* stand for, respectively, *internal* variables, *boundary* variables and the *remaining* variables. *Internal* variables are those local to the subcircuit; *boundary* variables are those related to both the subcircuit and the rest of the circuit. Note that boundary variables include those variables required as the circuit inputs and outputs. With this, the system-equation set $Ax = b$, can be rewritten in the following form:

$$\begin{bmatrix} A^{II} & A^{IB} & 0 \\ A^{BI} & A^{BB} & A^{BR} \\ 0 & A^{RB} & A^{RR} \end{bmatrix} \begin{bmatrix} x^I \\ x^B \\ x^R \end{bmatrix} = \begin{bmatrix} b^I \\ b^B \\ b^R \end{bmatrix}. \tag{5.1}$$

The matrix, A^{II}, is the *internal* matrix associated with the internal variable vector x^I. Assume that the row indices for A^{II} are from 1 to m, the row indices for A^{BB} are from $m+1$ to $m+l$, and the row indices for A^{RR} are from $m+l+1$ to n, where m and l are the sizes of the submatrices A^{II} and A^{BB} respectively.

Subcircuit suppression (also called Schur decomposition) is used to eliminate all the variables in x^I, and to transform (5.1) into the following reduced set of equations:

$$\begin{bmatrix} A^{BB*} & A^{BR} \\ A^{RB} & A^{RR} \end{bmatrix} \begin{bmatrix} x^B \\ x^R \end{bmatrix} = \begin{bmatrix} b^{B*} \\ b^R \end{bmatrix}, \tag{5.2}$$

where

$$A^{BB*} = A^{BB} - A^{BI}(A^{II})^{-1}A^{IB}, \tag{5.3}$$

$$= A^{BB} - \frac{1}{\det(A^{II})} A^{BI}[\Delta^{II}_{u,v}]^T A^{IB}, \tag{5.4}$$

where $[\Delta_{u,v}]^T$ is called the *adjoint* matrix of A, $\Delta_{u,v}$ is the first-order *cofactor*, Schur decomposition of $\det(A)$ with respect to $a_{u,v}$, and is defined as $\Delta_{u,v} = (-1)^{(u+v)}\det(A_{a_{u,v}})$, and matrix $A_{a_{u,v}}$ is the $(n-1) \times (n-1)$ matrix obtained from matrix A by deleting row u and column v. A^{BB*} is also called the Schur complement [40]. We also have

$$b^{B*} = b^B - A^{BI}(A^{II})^{-1}b^I, \tag{5.5}$$

$$= b^B - \frac{1}{\det(A^{II})} A^{BI}[\Delta^{II}_{u,v}]^T b^I. \tag{5.6}$$

Subcircuit suppression can be performed for all the subcircuits by visiting the circuit hierarchy in a bottom-up fashion. Since the number of internal variables is m, and the number of boundary variables is l, (5.4) and (5.6) can be written in the following expanded forms:

$$a^{BB*}_{u,v}(s) = a^{BB}_{u,v}(s) - \frac{1}{\det(A^{II}(s))} \sum_{k_1,k_2=1}^{m} a^{BI}_{u,k_1}(s)\Delta^{II}_{k_2,k_1}(s)a^{IB}_{k_2,v}(s), \tag{5.7}$$

where $u, v = m+1, ..., m+l$ and

$$b^{B*}_u(s) = b^B_u(s) - \frac{1}{\det(A^{II}(s))} \sum_{k_1,k_2=1}^{m} a^{BI}_{u,k_1}(s)\Delta^{II}_{k_2,k_1}(s)b^I_{k_2}(s), \tag{5.8}$$

where $u = m+1, ..., m+l$. From (5.7) and (5.8), we can observe that admittance $a^{BB*}_{u,v}$ and input stimuli b^{B*}_u at boundary nodes will become rational functions of s once the subcircuit is suppressed. The term

$$(a^{BI}_{u,k_1}(s)\Delta^{II}_{k_2,k_1}(s)a^{IB}_{k_2,v}(s))/\det(A^{II}) \tag{5.9}$$

is called the *composite admittances* as they are the admittances generated during subcircuit reduction.

To obtain the rational functions for $a^{BB*}_{u,v}(s)$ and $b^{B*}_u(s)$, we need to compute the rational function for a determinant whose elements may again be rational functions of s. This task can be achieved using determinant decision diagrams (DDDs), which were originally proposed for symbolic analysis of large analog circuits [97, 110, 126] and will be discussed in Section 5.2. We will show how the DDD graphs can be modified to compute the rational function for a determinant in later sections.

An important issue is that if the symbolic expressions are kept (s-expressions are special symbolic expressions), the final expressions of the generated rational admittances are not free from cancellation. There are two cancellation types. The first one is common-factor cancellation from common factors between the numerator and denominator of the resulting rational admittances in $a_{u,v}^{BB*}(s)$ and $b_u^{B*}(s)$. The second one is term cancellation from canceling terms (the sum of those symbolic terms equals to zero).

Such symbolic cancellation problems have been observed and discussed in [96,97] in the context of Y-Δ transformation. The cancellation of common factors may lead to exponential growth of the magnitude of coefficients in the numerators and denominators. As we can see in the later sections of this chapter, cancellation-free rational admittances are related directly to the exact admittance expressions computed from the flattened circuit matrix and are also very important to the new de-cancellation method.

Fundamentally, all the cancellations are caused by subcircuit reductions as shown in [122]. The detailed explanation of the conditions for common-factor cancellations and term cancellations can be found in [97,122].

Therefore, how to remove those common factors (de-cancellation) in the rational functions of $a_{u,v}^{BB*}(s)$ and $b_u^{B*}(s)$ in general becomes a key issue for the subcircuit reduction process. An efficient algorithm has been proposed for de-cancellation during the generalized Y-Δ transformation [97,122]. We will review how cancellation-free rational functions can be obtained by efficient DDD graph operations during the general hierarchical circuit reduction process.

5.2 DDD-based hierarchical decomposition

The idea of DDD-based hierarchical decomposition [126] is to represent all the determinants and cofactors in (5.7) and (5.8), as well as those determinants and cofactors in the final transfer functions of the desired circuit characteristics. Owing to the compactness of DDD graphs, DDD-based hierarchical circuit decomposition was shown to be superior to the previous methods [51,117].

5.2.1 *Determinant decision diagram*

In this subsection, we provide a brief overview of the notion of determinant decision diagrams [97,110].

Determinant decision diagrams [110] are compact and canonical graph-based representations of determinants. The concept is best illustrated using a simple RC filter circuit like the one shown in Figure 5.2.

Its system equations can be written as

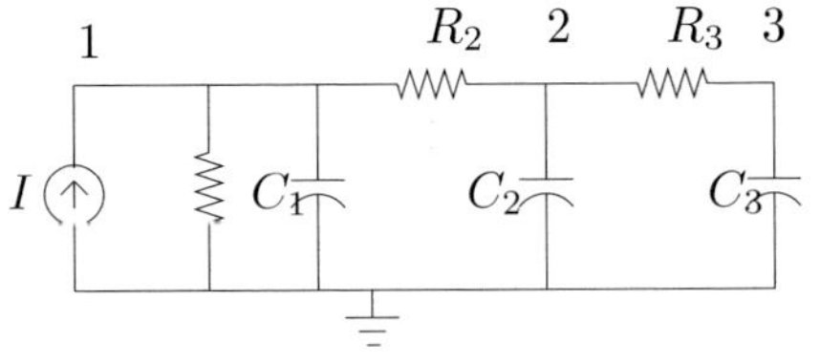

Figure 5.2 A simple RC circuit.

$$\begin{bmatrix} \frac{1}{R_1} + sC_1 + \frac{1}{R_2} & -\frac{1}{R_2} & 0 \\ -\frac{1}{R_2} & \frac{1}{R_2} + sC_2 + \frac{1}{R_3} & -\frac{1}{R_3} \\ 0 & -\frac{1}{R_3} & \frac{1}{R_3} + sC_3 \end{bmatrix} \begin{bmatrix} v_1 \\ v_2 \\ v_3 \end{bmatrix} = \begin{bmatrix} I \\ 0 \\ 0 \end{bmatrix}.$$

We view each entry in the circuit matrix as one distinct symbol, and rewrite its system determinant in the left-hand side of Figure 5.3. Then its DDD representation is shown in the right-hand side. Please refer to [110] for the formal definition of a DDD graph.

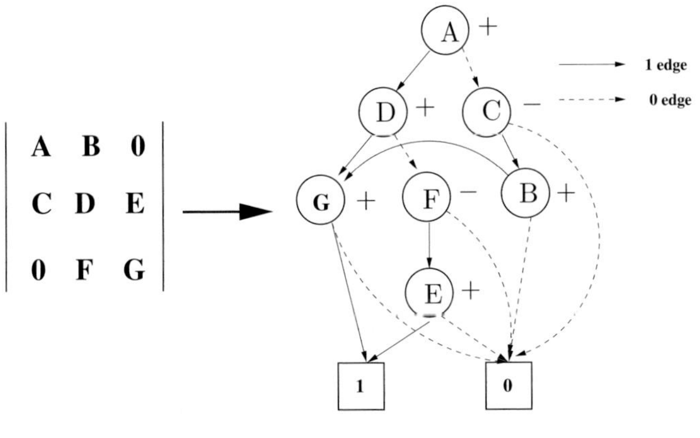

Figure 5.3 A matrix determinant and its DDD.

A DDD is a signed, rooted, directed acyclic graph with two terminal vertices, namely the *0-terminal* vertex and the *1-terminal* vertex. Each non-terminal DDD vertex is labeled by a symbol in the determinant, denoted by a_i (A to G in Figure 5.3), and a positive or negative sign, denoted by $s(a_i)$. It *originates* two outgoing edges, called *1-edge* and *0-edge*. Each vertex a_i represents a symbolic expression $D(a_i)$ defined recursively as follows: $D(a_i) = a_i\ s(a_i)\ D_{a_i} + D_{\overline{a}_i}$, where D_{a_i} and $D_{\overline{a}_i}$ represent, respectively, the symbolic expressions of the vertices pointed by the 1-edge and 0-edge of a_i. The 1-terminal vertex represents expression 1, whereas the 0-terminal vertex represents expression 0. For example, vertex E in Figure 5.3 represents expression E, and vertex F represents expression $-EF$, and vertex D represents expression $DG - FE$. We can also say that a DDD vertex D represents an expression defined the DDD subgraph rooted at D.

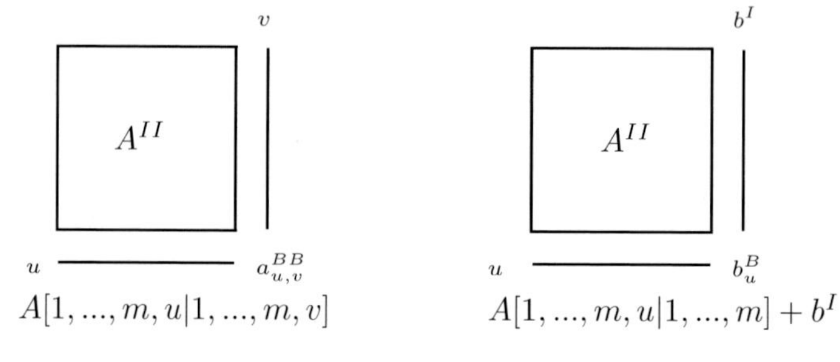

Figure 5.4 Illustration of Theorem 5.1. Reprinted with permission from [122] (c) 2005 IEEE.

A *1-path* in a DDD corresponds a product term in the original DDD, which is defined as a path from the root vertex (A in our example) to the 1-terminal including all symbolic symbols and signs of the vertices that originate all the 1-edges along the 1-path. In our example, there exist three 1-paths representing three product terms: ADG, $-AFE$ and $-CBG$. The root vertex represents the sum of these product terms. The size of a DDD is the number of DDD vertices, denoted by $|DDD|$. Note that the size of a DDD depends on the size of circuits in a very complicated way. $|DDD|$ is a linear function of the circuit size for ladder circuits and it may grow superlinearly in general with the sizes of circuits [110].

5.2.2 DDD representation for hierarchical subcircuit decomposition

First we introduce the following theorem: [124]

Theorem 5.1. (5.7) *can be written in the following form:*

$$a^{BB*}_{u,v} = \frac{\det(A[1, ..., m, u|1, ..., m, v])}{\det(A^{II})}, \tag{5.10}$$

where $u, v = m + 1, ..., m + l$ *and* (5.8) *can be written in the following form:*

$$b^{B*}_u = \frac{\det(A[1, ..., m, u|1, ..., m] + b^I)}{\det(A^{II})}, \tag{5.11}$$

where, $A[1, ..., m, u|1, ..., m, v]$ *is the submatrix that consists of matrix* A^{II}*, which actually is* $A[1, ..., m|1, ..., m]$*, plus row* u *and column* v *of matrix* A*;* $A[1, ..., m, u|1, ..., m] + b^I$ *is the submatrix that consists of matrix* A^{II} *plus row* u *of matrix* A *and the right hand side column* b^I*. The* $A[1, ..., m, u|1, ..., m, v]$*, and* $A[1, ..., m, u|1, ..., m] + b^I$ *are illustrated in Figure 5.4.*

With Theorem 5.1, we only need to compute the rational function for the numerator, which is a determinant (as $\det(A^{II})$ shared by all the newly created matrix elements for each subcircuit) for each boundary-variable related element in the reduced circuit matrix instead of representing each individual first-order cofactor of

$\det(A^{II})$ explicitly [126]. This leads to the efficient DDD-based hierarchical decomposition method. Such a determinant-only DDD representation for all the involved matrix elements is more compact than the previous method [126], as first-order cofactors are not explicitly represented. Therefore, more sharing of common terms becomes possible among those cofactors and elements in row u, columns v and b^I. More importantly, such DDD-only representation of new admittances is more amenable to removing cancellation during a subcircuit reduction process as shown in the next section [122].

5.2.3 Y-expanded DDDs

For our problem, some elements (in admittance forms) of a circuit matrix will become a rational function of s during the hierarchical subcircuit reduction process. As a result, the construction of the rational function for such a determinant will be different from the method used for s-expanded DDDs [111]. To efficiently compute the rational function for a determinant and handle cancellations, a new DDD graph, called a Y-parameter expanded DDD (YDDD), is first introduced. Y-expanded DDDs are also DDD graph where each DDD node represents a device admittance or a composite admittance, as shown in Figure 5.5.

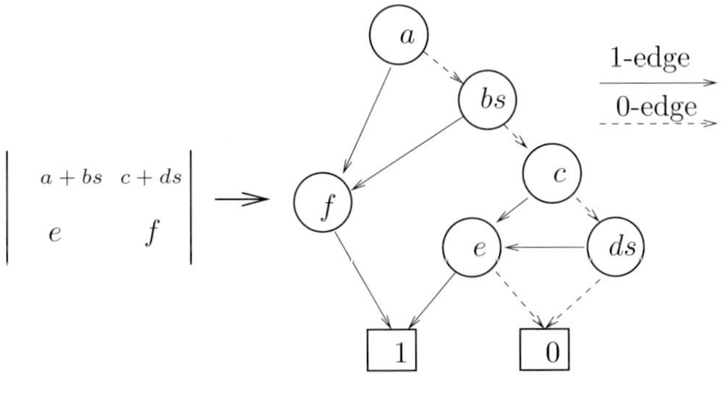

Figure 5.5 A determinant and its YDDD. Reprinted with permission from [122] (c) 2005 IEEE.

Note that some circuit parameter admittances are functions of the complex frequency variable s. This is different from the s-expanded DDD graphs where s is explicitly extracted and represented [111]. The main purpose of the introduction of YDDDs is that we can easily handle both term cancellation and common-factor cancellation as cancellation patterns can be easily detected by examining those device admittances or composite admittances. Like s-expanded DDDs [111], the YDDDs can be constructed from a complex DDD in linear time in the size of the original complex DDDs. The time complexity for constructing a complex DDD depends on the circuit topology. Given the best vertex ordering, if the underlying circuit is a ladder or tree circuit, $|DDD|$ is a linear function of the size of the circuit. For general circuits, the size of a DDD graph may grow exponentially in the worse case.

But like BDDs, with proper vertex ordering, the DDD representations are very compact for many real circuits [110, 111]. In Figure 5.2.3, we present a version of the algorithm called YDDDConst().

```
YDDDConst(D)
if ( D = 0 or D = 1)
  return NULL
L0 = YDDDConst(D.0)
L1 = YDDDConst(D.1)
P_result = NULL
for i = 1 to m do
  P_tmp = YDDDMultiply(L1,D.x_i)
  P_result = YDDDUnion(P_tmp, P_result)
return YDDDUnion(P_result, L0)
```

Figure 5.6 Y-expanded DDD construction. Reprinted with permission from [122] (c) 2005 IEEE.

Function YDDDConst() takes a complex DDD rooted at D and returns the resulting YDDD tree. *D.1* and *D.0* denote, respectively, the vertices pointed to by the 1-edge and 0-edge of vertex D. m is the number of devices admittance or composite admittances connected to a node, each of which is denoted by $D.x_i$. YDDDUnion(P_1, P_2) computes the union of two YDDDs, P_1 and P_2. YDDDMultiply(P, v) computes the product of YDDD P and YDDD vertex v.

It was shown that YDDD representation is more compact than the *sequence of expressions* [51, 126], whose complexity essentially represents the complexity of symbolic node-by-node Gaussian elimination process. Hence time complexity of the general hierarchical reduction algorithm is better than the Y-Δ reduction algorithm [96], which is based on Gaussian elimination. Another advantage of the general reduction algorithm over the existing Y-Δ transformation algorithm is that we need to remember a smaller number of common factors, which are the determinants of the reduced subcircuit matrices. If we reduce one node at a time, each node becomes a subcircuit, we end up with more common factors to deal with. Also there are more independent subcircuit hierarchies as each subcircuit is the smallest (consists of one node). As a result, we have to remember all the common factors of the reduced circuits (nodes) that are connected to nodes yet to be reduced, which in turn leads to more memory usage compared to the general reduction method. Our experimental results confirm those observations.

With YDDD, we can compute the s-polynomial of a determinant in a concellation-free manner. But in the context of hierarchical decomposition, we need to compute the rational admittance functions in a hierarchical and cancellation-free way with limited orders of s. The detail of this algorithm can be found at [97, 122].

For large RLC interconnect circuits, truncations have to be carried out to keep limited orders of s in each rational admittance. The resulting admittances or transfer functions may not be stable. A convex programming based method will be carried out to enforce the passivity of the reduced models [21], which will be discussed in Chapter 10.

5.2.4 Overview of passive hierarchical model order reduction (HMOR) method

In this subsection, we give the overview of the general frequency-domain hierarchical model order reduction and realization flow. We show how the symbolic hierarchical subcircuit decomposition and rational function computations can be used to perform the hierarchical model order reduction. The whole reduction flow algorithm consists of several important steps.

The key step for HMOR is hierarchical circuit reduction and model order reduction. The basic idea is to reduce subcircuits in a hierarchical and cancellation-free way. The new admittances coming from subcircuit suppressions shown in (5.10) and (5.11) are kept as rational functions of s in the exact or order-reduced form (with fixed order of s). Such a reduction can be repeated until we reach the top-level circuit, which is typically small enough to be solved exactly and symbolically (s is still the only symbol). The reduction process is repeated for multiple frequency (expanded at multiple frequency points) for wideband modeling. The resulting circuit unknown variables are rational functions of s, which can be further optimized to enforce passivity and be realized for easy use with SPICE-like circuit simulators.

The whole hierarchical model order reduction algorithm is described below:

THE GENERAL HIERARCHICAL NETWORK MODELING ALGORITHM

1. Build the complex DDDs for each subcircuit matrix and all the required cofactors and determinants for each newly created admittances in a bottom up way.
2. If the present circuit has subcircuits, perform the reduction on each subcircuit first.
3. Derive the YDDDs from complex DDDs for each new composite admittance shown in (5.10) and (5.11).
4. Construct the cancellation-free rational admittances for each YDDD.
5. Repeat step 1 to step 4 for multiple frequency points based on the given error bound.
6. Select the dominant poles from different frequency regions to form the final set of poles, which cover all the frequency range.
7. Apply convex programming passivity enforcement and optimization to determine all the residues of each rational admittance in the reduced admittance matrix (Chapter 10).
8. Realize the passivity-enforced order reduced admittance matrix using a general network synthesis technique (Chapter 11).

The general hierarchical reduction algorithm is also illustrated in Figure 5.7.

In the following sections, we first present some theortical results, which show that the one-point hierarchical model order reduction is equivalent to the implicit moment-matching around $s = 0$ and hierarchical model order reduction preserves the reciprotical property of a circuit. Then we present the multi-point hierarchical model order reduction scheme for wideband modeling.

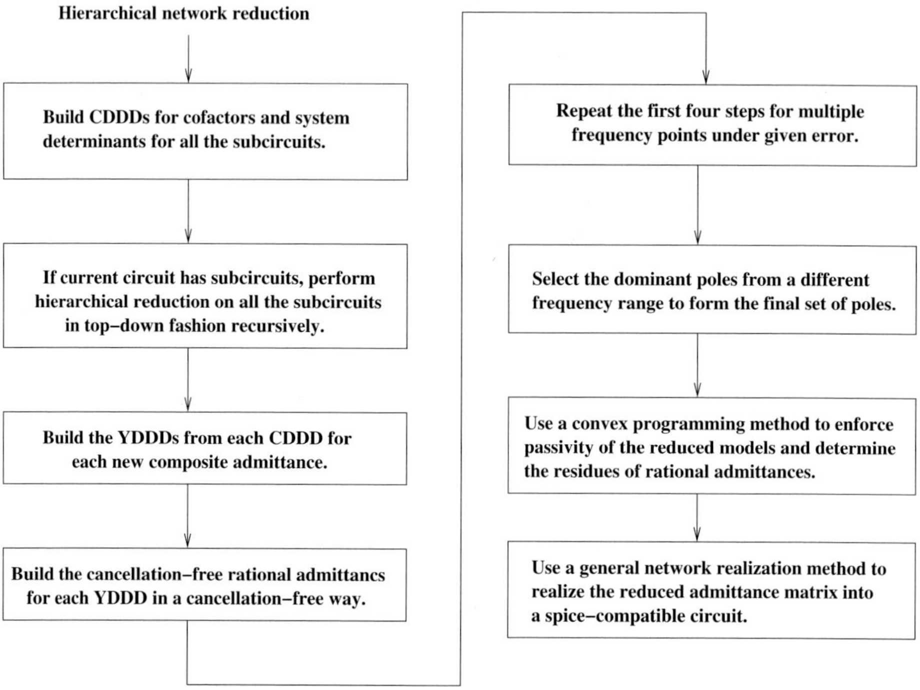

Figure 5.7 The general hierarchical model order algorithm flow.

5.3 Hierarchical reduction versus moment-matching

In this section, we first discuss how the frequency-domain hierarchical reduction is related to the implicit moment-matching. Then we discuss the numerical stability and reciprocity-preserving property of the hierarchical reduction process.

5.3.1 *moment-matching connection*

Consider a linear system with n state variables in vector x, the system is given by

$$sx = Ax + b, \tag{5.12}$$

where A is an $n \times n$ system matrix, b is the input vector to the circuit. Then we can obtain $x = (Is - A)^{-1}b$. Let us consider single-input single-output systems, where we have only one input b_j and we are interested in state response at node i. In this case we have

$$x_i(s) = H_{ij}(s)b_j = \frac{\Delta_{ij}}{\det(Is - A)}b_j, \tag{5.13}$$

where Δ_{ij} is the first-order cofactor of matrix $M = (Is - A)$ with respect to the element at the row i and column j and $H_{ij}(s)$ is the transfer function. So the exact solution of any state variable or its transfer function in frequency-domain can be represented by a rational function of s.

Hierarchical reduction, basically, is to reduce the $n \times n$ matrix M into a very smaller $m \times m$ matrix M' based on block Gaussian elimination, such that x_i can be trivially solved symbolically by using (5.13). During this reduction process, all the rational functions involved are truncated up to a fixed maximum order and the final solution will be a rational function with the same order for its numerator and its denominator. We then have the following theoretical result for the computed state variable $x'_i(s)$ from the hierarchical reduction process in the frequency-domain.

Theorem 5.2. *The state variable $x'_i(s)$ computed by the frequency-domain hierarchical reduction with q as the maximum order for all the rational functions will match the first q moments of the exact solution $x_i(s)$ expanded by Taylor series at $s = 0$.*

Proof: We know that the exact solution of $x_i(s)$ is a rational function as shown in (5.13). Because of the truncation, the solution computed by the hierarchical reduction process will be given by

$$x'_i(s) = \frac{a_0 + a_1 s + ... + a_q s^q}{b_0 + b_1 s + ... + b_q s^q}, i - 1, ..., q. \tag{5.14}$$

It was proved in [123] that a cancellation-free rational expression from the hierarchical reduction process is the exact expression obtained from the flat circuit matrix. If we do not perform any truncation, $x'_i(s)$ will be the exact solution, $x_i(s)$, which is obtained from the flat circuit matrix by (5.13) when all cancellations are removed numerically during the hierarchical reduction process (assuming that no numerical error is introduced). With truncation, all the coefficients $a_0, ..., a_q$ and $b_0, ..., b_q$ are still exactly the same as that in $x_i(s)$. If we compute the moments of $x'_i(s) = m_0 + m_1 s + ...$, the first q moments can be uniquely determined by the $2q$ coefficients $a_0, ..., a_q$ and $b_0, ..., b_q$:

$$m_i = \frac{a_i - \sum_{k+l=i,k\leq i,l\leq i,k\neq 0}^{i} b_k m_l}{b}. \tag{5.15}$$

If b_0 is not zero, b is simply b_0, otherwise b will be the first non-zero coefficient b^t and the first q moments become the coefficients of $s^{-t} (t > 0)$ to that of s^{-t+q}. So the theorem is proved. Hence the transfer function $H'_{ij}(s)$ will also match the exact one up to the first q moments. QED.

For a general multi-input and multi-output system, each element in the reduced $m \times m$ admittance matrix $M'(s)$ becomes a rational function [124]:

$$a_{u,v}^{BB*} = \frac{\det(M[1, ..., m, u|1, ..., m, v])}{\det(M^{II})}, \tag{5.16}$$

where, $M[1, ..., m, u|1, ..., m, v]$ is a matrix that consists of matrix M^{II}. It is $M[1, ..., m|1, ..., m]$, plus row u and column v of matrix M. Then we have the following result:

Corollary 5.1. *Each rational admittance function* $a'^{BB*}_{u,v}(s)$ *in the reduced* $m \times m$ *matrix,* $M'(s)$*, by the hierarchical reduction process, will match the first* q *moments of the exact rational function* $a^{BB*}_{u,v}(s)$ *expanded by Taylor series at* $s = 0$.

5.3.2 *Numerical stability of the hierarchical reduction*

The hierarchical reduction process is essentially equivalent to implicit moment-matching at $s = 0$. As a result, the frequency response far away from $s = 0$ will become less accurate owing to the truncation of high order terms. Another source of numerical error comes from the numerical de-cancellation process, where polynomial division is required for removing the common factors (cancellation) in the newly generated rational function, which will, in turn, introduce error from numerical term cancellation (the sum of two symbolic terms should have been zero, but is not zero owing to numerical error). Such numerical noise will cause the higher-order terms to be less accurate even if we try to keep them. In Figure 5.8, we show the responses from the 3-way, 2-level partitioned $\mu A741$ circuit [126] under different maximum reduction orders of rational functions. As we can see, increasing the rational function order does not increase the accuracy of the response after the order reaches eight. This is the typical numerical stability problem with the moment-

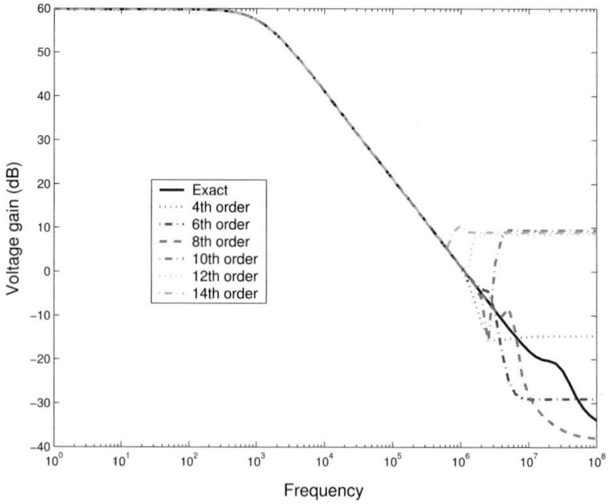

Figure 5.8 Frequency responses of $\mu A741$ circuit under different reduction orders. Reprinted with permission from [94] (c) 2006 IEEE.

matching method [91]. However, unlike explicit moment-matching methods, the hierarchical reduction is numerically stable for tree-structured circuits. We then show the following results:

Theorem 5.3. *For tree-structured circuits, the hierarchical reduction process can be performed such that there is no common-factor cancellation in the generated rational functions.*

Proof: For tree structured circuits, we can always partition the circuit in such a way that each subcircuit only has one node shared with its parent circuit. As a result, there is only one composite admittance $a^{BI}_{u,k_1}\Delta^{II}_{k_2,k_1}a^{IB}_{k_2,v}/\det(M^{II})$ in (5.7) generated in its parent circuit for each subcircuit. According to the common-factor cancellation condition [121], at least four composite elements from the same subcircuit reduction are required for the existence of common-factor cancellation. So no common-factor cancellation will occur under such partitioning. QED.

The significance of Theorem 5.3 is that the hierarchical reduction process becomes numerically stable for an arbitrary order for tree circuits. The only cancellation left is the term cancellation, where the sum of two symbolic terms is zero, which will not introduce any noticeable numerical error in the reduction process. Figure 5.9 shows the voltage gain response (real part) of an RC tree with about 100 nodes (also 100 capacitors) under different reduction orders. As can be seen, the reduced voltage gain will match the exact one well when the kept orders reach about 60.

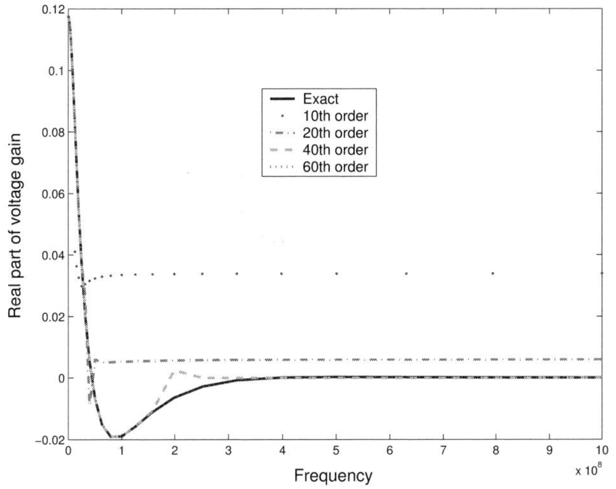

Figure 5.9 Frequency responses of an RC tree circuit under different reduction orders. Reprinted with permission from [94] (c) 2006 IEEE.

The fact that no common-factor de-cancellation (polynomial division) is required was also exploited in the direct truncation of the transfer-function method

(DTT) [56], where only polynomial addition is required to compute the truncated transfer functions for tree-structured RLC circuits. The DTT reduction process can be viewed as a special case of our method. But for general non-tree structured circuits, polynomial division is required in node-elimination-based reduction methods owing to common-factor cancellation, and polynomial division owing to truncation will not be numerically stable for very high frequency ranges far away from DC, as shown before.

To mitigate this problem, we propose using multi-point expansion to obtain accurate rational functions or reduced admittance matrices for modeling a general multi-input and multi-output linear system, as will be shown in Section 5.5.

5.4 Preservation of reciprocity

A reciprocal network is one in which the power losses are the same between any two ports regardless of the direction of propagation [127]. Mathematically, this is equivalent to the requirement that the circuit admittance matrix is symmetric (or scattering parameter $S21 = S12$, $S13 = S31$, etc). A network is known to be reciprocal if it is passive and contains only isotropic materials. Reciprocity is an important network property. For the hierarchical reduction, we have the following result:

Theorem 5.4. *The hierarchical reduction method preserves the reciprocity of a passive circuit matrix.*

Proof: The proof can be found by using (5.10) again. We first study a circuit with circuit matrix M. The circuit has one subcircuit with circuit matrix M^{II}. Assume that the original circuit matrix is symmetric (its subcircuit is also symmetric, i.e., both M and M^{II} are symmetric) owing to reciprocity.

After reduction, the reduced circuit matrix becomes an $m \times m$ matrix where each matrix element at row u and column v appears in (5.10). Then we look at the element at row v and column u, which is

$$a_{v,u}^{BB*} = \frac{\det(M[1, ..., m, v|1, ..., m, u])}{\det(M^{II})}, \tag{5.17}$$

Notice that matrix M is symmetric, so row u in (5.10) and column u in (5.17) are the same. This is true for column v in (5.10) and row v in (5.17). As a result we have

$$M[1, ..., m, v|1, ..., m, u] = M[1, ..., m, u|1, ..., m, v]^T, \tag{5.18}$$

$$\det(M[1, ..., m, v|1, ..., m, u]) = \det(M[1, ..., m, u|1, ..., m, v]^T). \tag{5.19}$$

Hence, $a_{u,v}^{BB*} = a_{v,u}^{BB*}$ and reciprocity is preserved in the reduced circuit matrix when a subcircuit is reduced. In the hierarchical reduction, we reduce one subcircuit at a time and the reduced circuit matrix is still symmetric after reduction. So the reduced circuit matrix is still symmetric after all the subcircuits are reduced. QED.

5.5 Multi-point expansion hierarchical reduction

The multi-point expansion scheme by real or complex frequency shifting has been exploited before in projection based reduction approaches for improving the modeling accuracy [19, 53]. The basic idea for such a strategy is that dominant poles close to an expansion point is more accurately captured than the poles that are far away from the expansion point in the moment-matching-based approximation framework. Therefore, instead of expanding at only one point, we can expand at multiple points along the real or complex axis to capture all the dominant points in the given frequency range accurately.

In this chapter, we extend this concept to the hierarchical reduction algorithm. Specifically, at each expansion point, the driving point function or each rational admittance function in a reduced admittance matrix can be written into the partial fraction form

$$f(s) = \sum_{i}^{n} k_i/(s - p_i). \tag{5.20}$$

By intelligently selecting poles and their correspond residues from different expansions and combining them into one rational function, we can obtain a more accurate rational function for a very high frequency range. In this chapter, we propose an explicit waveform matching scheme based on hierarchical reduction framework to find dominant poles and their residues for both SISO (single-input single-output) and MIMO (multi-input multi-output) systems. It is shown experimentally to be superior to the existing pole searching algorithm.

5.5.1 *Multi-point expansion in hierarchical reduction*

To expand the circuit at an arbitrary location in the complex s-plane, say $s_k = \alpha_k + \omega_k j$, we can simply substitute s in (5.13) by $s + s_k$. Then (5.13) becomes

$$x_i(s) = H_{ij}(s)b_j = \frac{\Delta_{ij}(s + s_k)}{\det(I(s + s_k) - A)} b_j. \tag{5.21}$$

As shown in [19], poles that dominate the transient response in interconnect circuits are near the imaginary axis with large residues. Hence, we expand along the imaginary axis for RF passive and interconnect circuits. Since only capacitors and inductors are associated with the complex frequency variable s, expansion at a real point α or a complex point $\omega_i j$ is essentially equivalent to analyzing a new circuit where each capacitor C has a new resistor (with real value $\alpha_i C$ or complex value $\omega_i C j$) connected in parallel with it and each inductor L has a new resistor (with real value $\alpha_i L$ or complex value $\omega_i Lj$) connected in series with it [92].

In this chapter, we show that the multi-point expansion can be made very efficiently in the hierarchical reduction framework. The rational functions are constructed in a bottom-up fashion in a Y-parameter determinant decision diagram (YDDD) in the hierarchical reduction algorithm [121]. When a capacitor C or an inductor L (its YDDD node) is visited, we build a simple polynomial $0 + Cs$ or $0 + Ls$ to multiply or add to existing polynomials seen at that DDD node. In the presence of a non-zero expansion point, α_i or $\omega_i j$, we can simply build a new polynomial $\alpha_i C + Cs$ or $\omega_i Cj + Cs$ for the capacitor and $\alpha_i L + Ls$ or $\omega_i Lj + Ls$ for the inductor respectively. So we do not need to rebuild the circuit matrix or the YDDD graphs used for reduction at $s = 0$. Instead, we only need to rebuild the rational functions by visiting every YDDD node once, which has the time complexity linear with the YDDD graph size, a typical time complexity for DDD-graph-based methods [110].

5.5.2 *Explicit waveform-matching algorithm*

One critical issue in multi-point expansion is to determine, at each expansion point, which poles are accurate and should be included in the final rational function. In the complex frequency hopping method [19], a binary search strategy was used where poles (common poles) seen by two expansion points and poles with distance to the expansion points shorter than the common poles are selected. Such a *common-pole matching* algorithm, however, is very sensitive to the numerical distance criteria for judging if two poles are actually a same pole. For accurately detecting common poles, a small distance is desirable, but this will lead to more expansion points; and even worse is that the same pole may be treated as different poles seen by two different expansion points. Also, this method may fail to detect some dominant poles as the circle for searching accurate poles might be too small, as shown in our experimental results.

In this chapter, we propose a new reliable pol-searching algorithm, which is based on explicit frequency waveform matching. The new algorithm is based on the observation that a complex pole p_i and its residue k_i in the partial fraction form, $k_i/(s - p_i)$, has the largest impact at frequency f_i when the imaginary part of the pole equals $2\pi f_i$. Figure 5.10 shows a typical response of $k_i/(s - p_i)$, where $k_i = 2.78 \times 10^{12} + 2.34 \times 10^{10} j$ and $p_i = -4.93 \times 10^8 + 2.58 \times 10^{10} j$. The peaks of both real (absolute) value and magnitude are around 4.11×10^9, which is equal to $2.58 \times 10^{10}/(2\pi)$. The reason for this is that both real and imaginary parts of $k_i/(s - p_i)$ reach a peak when their denominator $(pr)^2 + (\omega - pi)^2$ reaches a minimum at $\omega = pi$, where $s = \omega j$ and $p_i = pr + pij$. A complex pole with a negative imaginary part typically will not have significant impact on the upper half complex plane.

The idea of *frequency-waveform matching* is explicitly to match the approximate frequency waveform with that of exact ones. Specifically, at an expansion point, f_i, we perform the hierarchical reduction and then determine an accurate maximum frequency range $[f_i, f_{i+1}]$ such that the error between responses (magnitude) of the

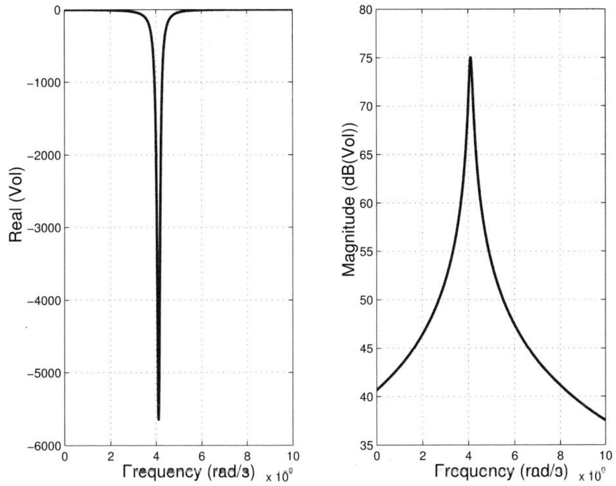

Figure 5.10 Responses of a typical $k_i/(s-p_i)$. Reprinted with permission from [94] (c) 2006 IEEE.

reduced rational function and that of the exact one are bounded by a pre-specified error bound. The error is computed as follows:

$$err = \frac{|\mathrm{dB20}(V_e) - \mathrm{dB20}(V_a)|}{|\mathrm{dB20}(V_e)|} \tag{5.22}$$

where $\mathrm{dB20}(x) = 20 * log_{10}(|x|)$ and $|x|$ is the magnitude of a complex number x; V_e is the exact response and V_a is the approximate response. If $|\mathrm{dB20}(V_e)| = 0$, then we use $|\mathrm{dB20}(V_a)|$ as the denominator in (5.22), if it is not zero. If $|\mathrm{dB20}(V_a)| = 0$, we have $err = 0$.

Then f_{i+1} will be the next expansion point. All the poles whose imaginary parts fall within the range $[2\pi f_i, 2\pi f_{i+1}]$ will be selected because their contribution in this frequency range is the largest. The new algorithm does not have the duplicate pole issue as accurate poles can only be located at one place. The accuracy of the found poles is assured by explicit waveform matching. Experimental results show that it tends to use less expansion points than the common-pole matching method, and less CPU time.

5.5.3 *Multi-point expansion for MIMO system reduction*

For a multi-input multi-output system, by using the modified nodal analysis, the reduced circuit matrix $M'(s) = [y_{ij}(s)]_{m\times m}$ will become an $m \times m$ admittance matrix. Each admittance y_{ij} is a complex rational function with real or complex (if expansion points are on the imaginary axis) coefficients. In this case, we explicitly watch for the error between each approximate rational admittance and the exact value of the admittance at each frequency. The exact value of each admittance can be computed by visiting the DDD graph representing the admittance. Since there is a lot of sharing among those admittances, the costs of evaluating all the admittances are similar to evaluating one admittance, considering that every DDD node just needs to be visited once at each frequency point [126].

5.6 Numerical examples

In this section, we present some numerical examples by using the HMOR methods and compare it with several other methods.

5.6.1 *One-port macro-model for spiral inductor*

We first construct a detailed PEEC model for a three-turn spiral inductor with its substrate. We assume copper ($\rho = 1.7 \times 10^{-8}\Omega \cdot m$) for the metal and the low-k dielectric ($\epsilon = 2.0$). The substrate is modeled as a lossy ground plane (heavily doped) with $\rho = 1.0 \times 10^{-5}\Omega \cdot m$. The conductor is volume discretized according to the skin depth, and longitudinally segmented by one 10th of the wave-length. The substrate is also discretized as in [79]. The capacitance is extracted by FastCap [83] and only adjacent capacitive coupling is considered, since capacitive coupling is short-range. The partial inductance is extracted by FastHenry at 50GHz [59]. The inductive coupling between any pair of segments (including segments in the same line) is considered. Then we generate the distributed PEEC model by π-type of RLC-topology to connect each segment, and it results in a SPICE netlist with 232 passive RLCM elements. The substrate parasitic contribution (Eddy current loss) is lumped into the above conductor segment. Note that for more accurate extraction at ultra-high frequencies, we need a full-wave PEEC model description [61]. For mutual inductance, a vector potential equivalent model (VPEC) is used [134], which is more hierarchical-reduction friendly as no coupling inductor branch currents are involved and circuit partition can be done easily.

5.6.2 *Comparison with common-pole matching method in frequency domain*

For the spiral inductor, the driving point impedance is obtained by the multi-point hierarchical reduction process. We use both the common-pole matching algorithm in the complex frequency-hopping method and the new waveform-matching algorithm to search for dominant poles along the imaginary axis.

For a fair comparison, we make sure that the resulting rational functions will have similar accuracy. For a common-pole matching algorithm, if two poles are located within 1% of their magnitude, they are regarded as the same pole. For the waveform-matching algorithm, the error bound between the approximate one and the exact one is set to 0.1%. As a result, the common-pole approach takes 26 expansions with 37.1 seconds, waveform marching method use 15 expansion with 22.57 seconds. The responses obtained using both methods versus the exact response up to 100GHz are shown in Figure 5.11. The responses from both methods match the exact ones very well all the way to 100 GHz. Our experience shows that the CPU time of the common-pole method is highly dependent on the common-pole detection criteria. For instance, if we set the criteria for common-pole detection to 0.5%, then 65 expansions are carried out. Also as more expansions are carried out, the chance that a single pole is seen by two consecutive expansion points becomes larger, but it may be treated as different poles, owing to a small distance criteria, which in turn leads to a significant distortion of the frequency response.

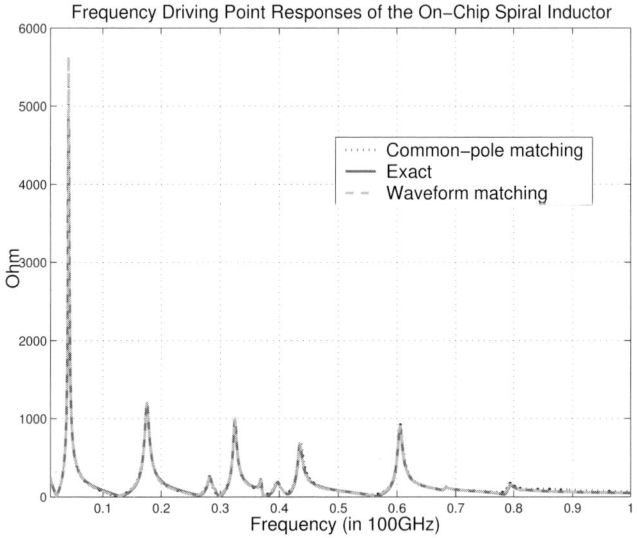

Figure 5.11 Frequency responses of the three-turn spiral inductor and its reduced model by using waveform matching and the common-pole method. Reprinted with permission from [94] (c) 2006 IEEE.

5.6.3 *Time-domain simulation of an LC oscillator*

We further demonstrate the accuracy and efficiency of the inductor macro-model in the time-domain harmonic simulation. Note that the synthesized one-port macro-model can be used to efficiently predict the critical performance parameters of spiral inductor, such as the ω_T, Q factor, and even the resonance starting-condition for an oscillator [99].

We use the Colpitts LC oscillator as an example, as shown in Figure 5.12 (b), where the active circuit behaves like a *negative resistor* to make the oscillator work, as shown in Figure 5.12 (a).

In this experiment, the synthesized one-port model is from a 25-order rational function with 24 poles and results in a macro-model with 40 RLC elements. As shown in Figure 5.13, the waveform in the steady state of the synthesized and original models match very well while we observe a 10× (5.17s as against 0.52s) runtime speed-up by using the reduced model.

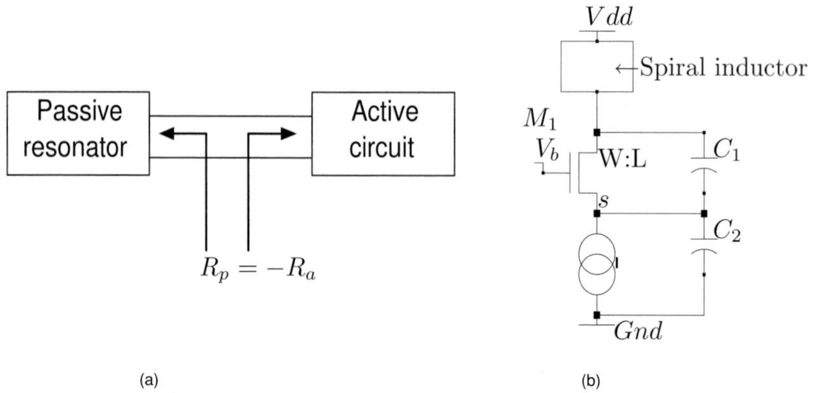

Figure 5.12 Colpitts LC oscillator with spiral inductors. Reprinted with permission from [94] (c) 2006 IEEE.

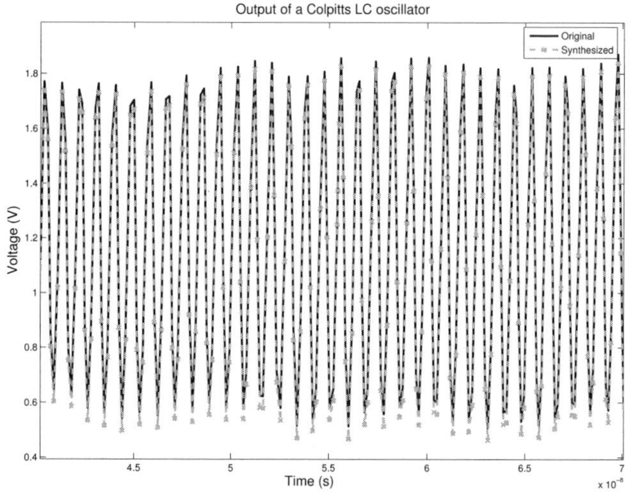

Figure 5.13 Time-domain comparison between original and synthesized models for a Colpitts LC oscillator with a three-turn spiral inductor. Reprinted with permission from [94] (c) 2006 IEEE.

5.6.4 Multi-port macro-model of coupled transmission line

We then use a two-bit coupled transmission line as the example for multi-port reduction and synthesis. The original PEEC model contains 42 resistors, 63 capacitors, 40 self-inductors, and 760 mutual inductors, where we consider inductive coupling between any two segments including those in the same line. Still, for mutual inductance, a vector potential equivalent model (VPEC) is used.

The matching frequency is up to 31 GHz and we find 24 dominant poles in

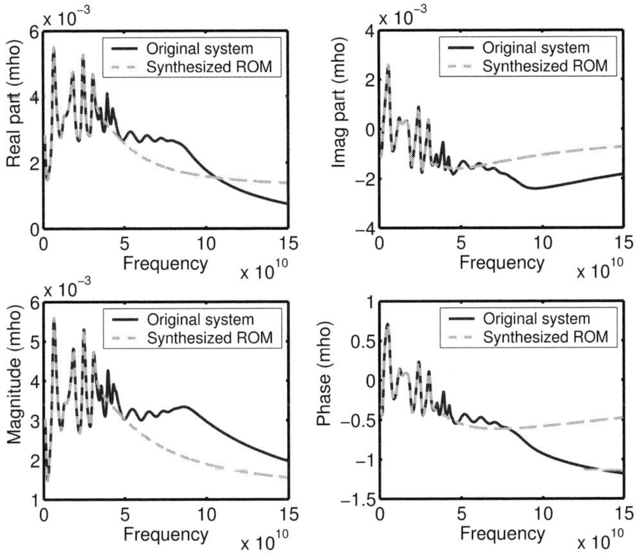

Figure 5.14 Frequency responses of Y_{11} of a two-bit transmission line. Reprinted with permission from [94] (c) 2006 IEEE.

this range. There are 150 RLC elements in the synthesized circuit compared with 364 devices in the original circuit; this represents a 58.79% reduction rate. The frequency responses for $Y_{11}(s)$ and $Y_{12}(s)$ are shown in Figure 5.14 and Figure 5.15, respectively. If we only match to 14 Ghz, 12 poles are required and we can achieve a 78.5% reduction rate instead. The time-domain step responses from the original circuit, the 14 GHz synthesized circuit, and the 31 GHz synthesized circuit are shown in Figure 5.16. The difference among these three circuits is fairly small.

In Figure 5.17, we further compare the frequency responses of the 24th-order macro-model by hierarchical model order reduction (HMOR), the 24th-order macro-model by PRIMA, and time-constant-based reduction (with similar reduction ratios) with the original circuit. The frequency response of port one is observed at the input port of the first bit, and that of port two is at the far end of the first bit. Owing to the preserved reciprocity, the reduced mode is easily realized by Foster's synthesis, and the model size is half of the SPICE-compatible circuit by PRIMA (via recursive convolution for each Y_{ij}). Moreover, as shown in Figure 5.17, the accuracy of the 24th-order model by HMOR can match up to 30 GHz but the same order model by PRIMA can only match up to 20 GHz. Note that under the similar reduction ratio with HMOR, the time-constant-based reduction can only match up to 5 GHz.

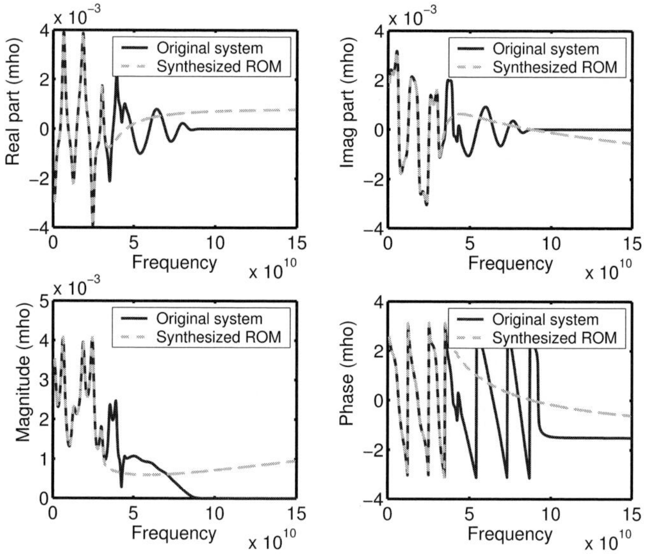

Figure 5.15 Frequency responses of Y_{12} of a two-bit transmission line. Reprinted with permission from [94] (c) 2006 IEEE.

1	2	3	4	5	6	7	8	9	10	11
Circuits	No. of	No. of	Simulation time (s)				Model size (kb)			
	elements	poles	HMOR	time const	PRIMA	Original	HMOR	time const	PRIMA	Original
circuit1	84	10	0.15	0.12	0.50	0.51	0.85	0.86	3.51	3.52
circuit2	258	10	0.15	0.15	0.92	5.31	0.85	0.86	3.51	11.23
circuit3	905	25	0.32	0.43	2.25	19.51	1.68	1.70	19.8	41.1
circuit4	5255	30	0.52	0.89	3.13	661.46	1.92	2.23	28.2	243.6
circuit5	20505	30	0.52	1.08	5.98	1356.66	1.92	2.53	28.2	957.9

Table 5.1 *Simulation efficiency comparison between original and synthesized model (part I). Reprinted with permission from [94] (c) 2006 IEEE.*

5.6.5 *Scalability comparison with existing methods*

Table 5.6.5 gives the reduction CPU time comparison for two methods; *HMOR* denotes the CPU times of hierarchical reduction. We notice that the HMOR is slow than the PRIMA. But the difference become less for large circuits. Theoretically, PRIMA and the one-point hierarchical model reduction have the same time complexity, that is the time of complexity of one Gaussian elimination. In PRIMA, we have to solve the circuit matrix at least once using Gaussian elimination or LU decomposition to solve for all the Krylov space base vectors (or moments). In the HMOR method, if we reduce one node at a time, it becomes a Gaussian elimination process. All the polynomial operations with fixed order have fixed computing costs.

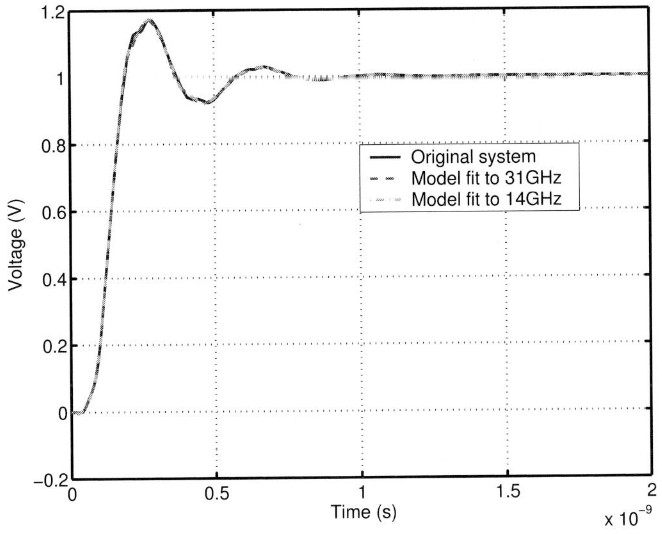

Figure 5.16 Transient responses of a two-bit transmission line. Reprinted with permission from [94] (c) 2006 IEEE.

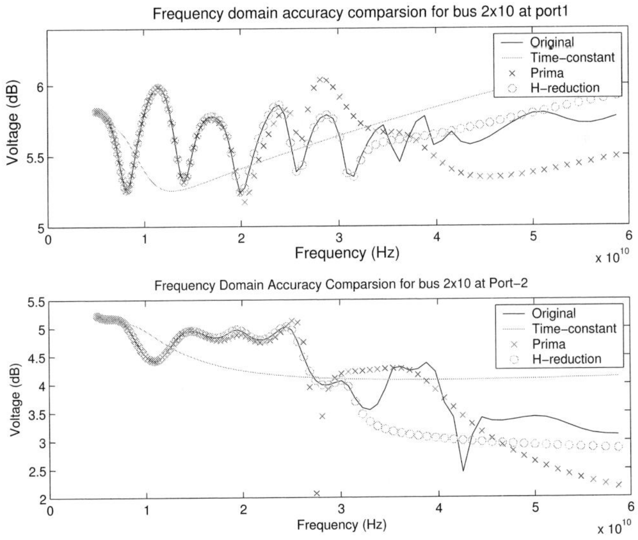

Figure 5.17 Frequency responses of a two-bit transmission line at two ports. Reprinted with permission from [94] (c) 2006 IEEE.

The efficiency difference is mainly owing to expensive recursive operations used in graph operations, which can be further improved, and multi-point matching. However the multi-point matching makes our method closed loop and this gives us good control on the model accuracy. For PRIMA, the model accuracy cannot be

Circuits	No. of elements	No. of poles	Delay error (%)		
			HMOR	time const.	PRIMA
circuit1	84	10	−0.16%	−0.86%	−0.15%
circuit2	258	10	−0.24%	−1.12%	−0.22%
circuit3	905	25	−0.41%	−4.43%	−0.37%
circuit4	5255	30	−0.62%	−6.83%	−0.58%
circuit5	20505	30	−1.04%	−12.91%	−0.83%

Table 5.2 *Simulation efficiency comparison between original and synthesized model (part-II). Reprinted with permission from [94] (c) 2006 IEEE.*

Circuits	No. of elements	PRIMA (s)	HMOR (s)
circuit1	84	0.042	0.6
circuit2	258	0.077	11.7
circuit3	905	0.16	17.8
circuit4	5255	1.73	44.4
circuit5	20505	32.37	96.8

Table 5.3 *Comparison of reduction CPU times. Reprinted with permission from [94] (c) 2006 IEEE.*

determined without several trials using different reduction orders.

We finally present a scalability comparison in Tables 5.1 and 5.2 by time-domain transient simulation for the following aspects: (i) runtime of simulation; (ii) realization efficiency (realized model size); and (iii) accuracy in terms of delay. Several different sized RLCM circuits are used. We compare our method (*HMOR*) with the time-constant-based circuit-reduction [3] (*time const.*) and projection-based reduction PRIMA implemented at [64] (*PRIMA*). The same number of poles is used for the reduction when we compare our HMOR with PRIMA. The reduced model by time-constant reduction is obtained with a similar model size as HMOR.

First, we find that the realized RLCM circuit model size is up to 10 times smaller on average than the SPICE compatible circuit from PRIMA. Therefore, a similar simulation speed-up ($8\times$) is observed when we run both circuits in SPICE3. When we further compare the simulation time of our reduced models with the PEEC circuits, a significant speed-up (up to $2712\times$ for circuit5) is obtained. Furthermore, the waveform accuracy in terms of delay is given in columns 4–6 in Table 5.2. The reduced models are very accurate with the worst case delay error being -1.04% even with $478\times$ ($957.9kb$ versus $1.92kb$) reduction ratio in terms of model size. But for the same reduction ratio as our reduction, we find that the time-constant-based reduction introduces large errors (up to 12.91%) because too many nodes are eliminated and the reduction criteria cannot be satisfied.

Note that the sparsification in the VPEC model can dramatically reduce the number of mutual inductive couplings, but can also maintain the accuracy [134]. As a result, we use this technique during our reduction for larger circuits. For example, in the case of circuit5 (the largest one) in Table 5.1 and 5.2, we obtain a 97.5% sparsification from 19900 to 498 mutual inductors. Because of this sparsification, the reduction time reduces by 10× (365.4s versus 47.8s).

5.7 Summary

In this chpater, we presented a new hierarchical multi-point reduction algorithm for wideband modeling of high-performance RF passive and linear(ized) analog circuits. On the theoretical side, we showed that the frequency-domain hierarchical reduction is equivalent to the implicit moment-matching around $s = 0$ and showed that the hierarchical one-point reduction is numerically stable for general tree-structured circuits. We also showed that the hierarchical reduction preserves the reciprocity of passive circuit matrices.

We presented a hierarchical multi-point reduction scheme for high-fidelity, wideband modeling of general passive and active linear circuits. A novel explicit waveform matching algorithm is presented for searching dominant poles and their residues from different expansion points; this is shown to be more efficient than the existing pole-search algorithm. The passivity of reduced models is enforced by state-space-based optimization method. We also presented a general multi-port network realization framework to generate SPICE-compatible circuits as the macro models of the reduced circuit admittance matrices. The resulting modeling algorithm can generate the multi-port passive SPICE-compatible model for any linear passive networks with easily controlled model accuracy and complexity. Experimental results on a number of passive RF and interconnect circuits have shown that the HMOR modeling technique generates more compact models given the same accuracy requirements than existing approaches, such as PRIMA and time-constant methods.

5.8 Historical notes on node-elimination-based reduction methods

As VLSI technology advances with increased operating frequency and decreased feature size, parasitics from on-chip interconnects and off-chip packaging will detune the performance of high-speed circuits in terms of slew rate, phase margin and bandwidth [2]. Reduction of design complexity especially for those extracted high-order RLCM network is important for efficient VLSI design verification.

Compact modeling of passive RLC interconnect networks has been a research intensive area in the past decade owing to increasing signal integrity effects and interconnect-dominant delay in current system-on-a-chip (SoC) design [72]. Existing approaches can be classified into two categories. The first category is based on subspace projection [32, 37, 63, 85, 91, 113]. The projection-based method was pioneered by asymptotic waveform evaluation (AWE) algorithm [91], where explicit moment-matching was used to compute dominant poles at low frequency. The Pade via Lanczos (PVL) [32], Arnoldi transformation [113] methods improved the numerical stability of AWE; congruence transformation method [63] and PRIMA [85] can further produce passive models. However, reduced circuit matrices by PRIMA are larger than direct pole marching (having more poles than necessary) [1] and PRIMA does not preserve certain important circuit properties, such as reciprocity [37]. The latest development by structured projection can preserve reciprocity [37], but it does not realize the reduced circuit matrices. An efficient reduction algorithm that is based on matching a few order moments and on realization is proposed in [62]. In general, no systematic approach has been proposed for realizing order-reduced circuit matrices.

Another quite different approach to circuit complexity reduction is by means of local node elimination and realization [3, 27, 98, 107, 108]. The major advantage of these methods over projection-based methods is that the reduction can be done in a local manner and no overall solution of the entire circuit is required so that reduced models can be easily realized using RLCM elements. This idea was first explored by selective node elimination for RC circuits [27, 107], where time-constant analysis is used to select nodes for elimination. Node reduction for magnetic coupling interconnect (RLCM) circuits has recently become an active research area. Generalized Y-Δ transformation [98], RLCK circuit crunching [3], and branch merging [108] have been developed, based on nodal analysis (NA), where inductance becomes susceptance in the admittance matrix. Since mutual inductance is coupled via branch currents, to perform nodal reduction, an equivalent six-susceptance NA model is introduced in [98] to reduce two coupling current variables and template matching via geometrical programming is used to realize the model order reduced admittances, but its accuracy depends heavily on the selection of templates and only 1-port realization has been reported. Meanwhile, RLCK circuit crunching and branch merging methods are first-order approximation based on the nodal time-constant analysis. The drawbacks for this first-order approximation are: (1) error is controlled in a local manner and will be accumulated, hence it is difficult to control the global error due to reduction; (2) not too many nodes can be reduced if the elimination condition is not satisfied.

Another way to model and characterize complex interconnect structures in high frequency (in RF or even microwave ranges) is by means of rational approximation based on direct measurements or rigorous full-wave electromagnetic simulation [1, 21, 22, 46, 49, 82, 105]. Many of those methods have been used in RF and microwave circuit modeling, as they are very flexible and can be applied to different interconnect structures and wideband modeling.

6 Terminal reduction of linear dynamic circuits

Complexity reduction and compact modeling of interconnect networks have been an intensive research area in the past decade, owing to increasing signal integrity effects and rising electro and magnetic couplings modeled by parasitic capacitors and inductors. Most previous research works mainly focus on the reduction of internal circuitry by various reduction techniques. The most popular one is based on subspace projection [32,37,85,91,113]. The projection-based method was pioneered by asymptotic waveform evaluation (AWE) algorithm [91], where explicit moment matching was used to compute dominant poles at low frequency. Later on, more numerical stable techniques were proposed [32,37,85,113] by using implicit moment matching and congruence transformation.

However, nearly all existing model order reduction techniques are restricted to suppress the internal nodes of a circuit. Terminal reduction, however, is less investigated for compact modeling of interconnect circuits. Terminal reduction is to reduce the number of terminals of a given circuit under the assumption that some terminals are similar in terms of performance metrics like timing or delays. Such reduction will lead to some accuracy loss. But terminal reduction can lead to more compact models after traditional model order reduction has been applied to the terminal-reduced circuit, as shown in Figure 6.1.

For instance, if we use subspace projection methods like PRIMA [85] for the model order reduction, a smaller terminal count will lead to smaller reduced models, given the same order of block moment requirement for both circuits. The reason is that for every block moment order increase, PRIMA will generate m new poles in the reduced models where m is the number of terminals. Hence, fewer terminals will directly lead to fewer poles used in the reduced models.

For terminal reduction, another question one may have is whether there are many similar terminals in many practical interconnect circuits? One important observation is that many terminals in practical interconnect circuits are close to each other structurally or are extracted by using mathematic discretization by volume or surface meshing in methods such as finite difference and finite element scheme. As a result, their electrical characteristics are also similar in a well-designed VLSI system. For instance, clock sinks in clock networks, substrate plane, and critical interconnects in the memory circuits like word or bit lines are among those interconnects. Recent studies [31,34] show that there exists a large degree of correlation

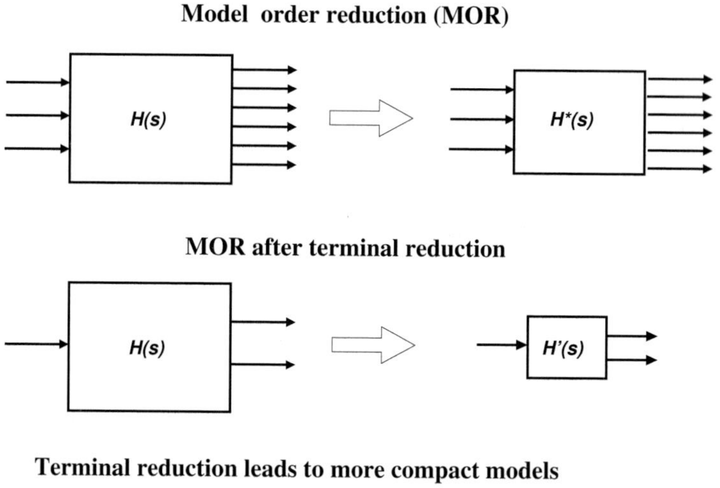

Figure 6.1 Terminal reduction versus traditional model order reduction.

between the various input and output terminals. A terminal reduction method named, SVDMOR was proposed [31,34]. The method, which is based on a low-rank approximation, was performed on the input and output position matrices before the model-order-reduction process. However, the low-rank approximation in the existing methods is only based on the DC or a specific order of moments of responses. Hence, the port-reduced systems may not correlate well with the original systems in terms of timing or delay.

In this chapter, we present a novel terminal reduction method called *TermMerg*. The new terminal reduction method is based on the observation that if we allow some delay tolerance or variations, which actually cannot be avoided in today's VLSI chip manufacture and working environments, some of the terminals with similar timing responses can actually be suppressed or merged into one representative terminal during the reduction without affecting their modeling functionality. In contrast to the existing terminal reduction methods, the new approach uses high order moments as timing and delay metrics. Specifically, given some delay tolerances or variations, *TermMerg* employs a singular value decomposition (SVD) method to determine the number of clusters based on the low-rank approximation. Then the K-means clustering algorithm is used to group the moments of the terminals into different clusters. After the clustering, we pick one terminal that could best represent other terminals for each cluster. The new method can work with any passive model order reduction and ensure the passive models. In contrast, we show that SVDMOR does not generate passive models in general. Passivity enforcement in SVDMOR will significantly hamper the terminal reduction qualities. We also present an improved version of the SVDMOR method, called *ESVDMOR*, which improves the computation efficiency of SVDMOR as well as accuracy for the similar reduced model sizes when the used moment matrix does not give a good terminal correlation.

6.1 Review of the SVDMOR method

In this section, we briefly review the SVDMOR method for terminal reduction, which was proposed recently for reducing the terminals of interconnect circuits [31, 34].

For a linear RLC interconnect network with p input and q output terminals, we can apply modified nodal analysis to formulate it into the state space equation form

$$\begin{aligned} G\mathbf{x}(t) + C\dot{\mathbf{x}}(t) &= B\mathbf{u}(t) \\ \mathbf{y}(t) &= L\mathbf{x}(t), \end{aligned} \tag{6.1}$$

where $G \in R^{n\times n}$ and $C \in R^{n\times n}$ are the conductive and storage element matrices. $L \in R^{q\times n}$ and $B \in R^{n\times p}$ are the output and input position matrices. $\mathbf{y}(t) \in R^q$, $\mathbf{u}(t) \in R^p$. State variables $\mathbf{x}(t) \in R^n$ can be nodal voltages or branch currents of the linear circuit.

The circuit transfer function is

$$H(s) = L(G + Cs)^{-1}B. \tag{6.2}$$

Then the ith block moment of the system is defined as

$$\mathbf{m}_i = L(-G^{-1}C)^i G^{-1}B, \tag{6.3}$$

which is a $q \times p$ matrix function.

The block moment $\mathbf{m}_i$ can be directly computed in a recursive way

$$\begin{aligned} \mathbf{x}_0 &= G^{-1}B;\ \mathbf{m}_0 &&= L\mathbf{x}_0 \\ \mathbf{x}_1 &= -G^{-1}C\mathbf{x}_0;\ \mathbf{m}_1 &&= L\mathbf{x}_1 \\ \cdots \\ \mathbf{x}_i &= -G^{-1}C\mathbf{x}_{i-1};\ \mathbf{m}_i &&= L\mathbf{x}_i \text{ for } i > 0, \end{aligned} \tag{6.4}$$

The SVDMOR method exploits the fact that many terminals are not independent in terms of their timing information, which can be reflected in their frequency domain moments. As a result, we can perform the singular value decomposition (SVD) on a block moment of specific order. For instance, if we perform the SVD on the 0th order block moment (DC response) $\mathbf{m}_0$, we have

$$\mathbf{m}_0 = LG^{-1}B = U\Sigma V^T, \tag{6.5}$$

where U and V are orthogonal matrices and Σ is a diagonal matrix with singular values in the diagonal in a decreasing order. If there are k dominant singular values, we can use a k-rank matrix (a $k \times k$ full rank matrix) to approximate the original $\mathbf{m}_0$ based on the SVD theory as

$$\mathbf{m}_0 = U\Sigma V^T \approx U_k \Sigma_k V_k^T. \tag{6.6}$$

Notice that U_k is $q \times k$ matrix and V_k^T is a $k \times p$ matrix and Σ_k is a $k \times k$ matrix. After this, we can have the following expressions

$$B = B_b V_k^T, \tag{6.7}$$
$$L^T = L_c U_k^T, \tag{6.8}$$

where $B_b \in R^{n \times k}$ and $L_c \in R^{n \times k}$ are obtained using the Moore–Penrose pseudoinverse of V_k.

$$B_b = BV_k(V_k^T V_k)^{-1}, \tag{6.9}$$
$$L_c = L^T U_k (U_k^T U_k)^{-1}. \tag{6.10}$$

The circuit transfer function now becomes

$$H(s) = U_k L_c^T (G + Cs)^{-1} B_b V_k^T. \tag{6.11}$$

Notice that the transfer function $H_r(s)$,

$$H_r(s) = L_c^T (G + Cs)^{-1} B_b, \tag{6.12}$$

which is inside (6.11), is a $k \times k$ matrix transfer function, which actually is the terminal-reduced transfer function of (6.2) and can be reduced by traditional Krylov subspace-based model order reduction methods. If the reduced transfer function of (6.12) is $\hat{H}_r(s)$, then the final order reduced transfer function is

$$\hat{H}(s) = U_k \hat{H}_r(s) V_k^T. \tag{6.13}$$

The SVDMOR method has several limitations. First, when the input and output terminals are quite different (especially in terms of numbers), SVDMOR does not work very well as SVDMOR performs the terminal reduction on both input and output responses at the same time. As a result, it can only approximate well for one type of terminals which have smaller terminal counts (as the rank of $\mathbf{m_i}$ typically is determined by the smaller dimensions of its rows or columns). We will compare SVDMOR with the improved SVDMOR, called ESVDMOR in Section 6.3.

The second problem with SVDMOR is that it is difficult to enforce the passivity during the combined terminal and model order reduction, and passivity terminal reduction of SVDMOR leads to less effect terminal reduction. We will discuss the passivity issues in Subsection 6.6.4.

6.2 Input and output moment matrices

Our task is to find the terminals with similar delay or timing behaviors such that they can be viewed as one terminal if some delay uncertainty is allowed. We focus on the timing metric of terminals and look at their timing responses from step or impulse inputs, but the new methods can be applied to other metrics of interest.

Ideally, the delay or timing information should be represented by waveforms in the time domain. However, this does not give us the best representation for the terminal merging method. Because all the waveforms have to be computed first, it is also difficult to compare two transient waveforms in general. Instead, we use terminal response moments in the frequency domain to represent their time-domain response information.

The terminal merging algorithm is based on the observation that if two terminals have similar timing or delay, then they should have similar moments (vectors) numerically. Remember that the moments computed in (6.17) represent the impulse responses of outputs from inputs. It is well known that the 1th moment m_1 represents the first-order delay approximation, or the Elmore delay, of the corresponding output with respect to a specific input. Higher-order moments represent more detailed timing and delay information. Hence, the moment vector is an ideal expression of timing information for the terminal merging algorithm.

Since we need to merge both input and output terminals, we need to present the timing information for both input and output terminal. As a result, we have two moment matrices, the input moment matrix M_I and the output moment matrix M_O.

For a 1×1 system, the system's transfer function $H(s)$ could be expanded into a Taylor series around $s = 0$. The coefficient of s^i in the series expansion is the ith moment of the transfer function:

$$H(s) = m_0 + m_1 s + m_2 s^2 + m_3 s^3 + \ldots \tag{6.14}$$

For a general linear (linearized) time-invariant network with p input and q output terminals, we can describe it as follows.

$$\begin{aligned} \dot{\mathbf{x}} &= A\mathbf{x} + B\mathbf{u}, \\ \mathbf{y} &= C\mathbf{x}, \end{aligned} \tag{6.15}$$

where $\mathbf{x}$ is an n-dimensional state vector, $\mathbf{u}$ is a p-dimensional input vector, and $\mathbf{y}$ is a q-dimensional output vector. The transfer function by Laplace transformation is

$$H(s) = C(sI - A)^{-1}B. \tag{6.16}$$

If we expand the above equation in a Taylor's series at $s = 0$, we get the moments at various terminals:

$$\begin{aligned} \mathbf{m}_0 &= -CA^{-1}B, \\ \mathbf{m}_1 &= -CA^{-2}B, \\ &\vdots \\ \mathbf{m}_{r-1} &= -CA^{-r}B. \\ &\vdots \end{aligned} \tag{6.17}$$

Each moment $\mathbf{m}_i$ is a $q \times p$ matrix,

$$\mathbf{m}_i = \begin{bmatrix} m^i_{1,1} \; m^i_{1,2} \cdots m^i_{1,p}, \\ m^i_{2,1} \; m^i_{2,2} \cdots m^i_{2,p}, \\ \vdots \quad \vdots \quad \vdots \quad \vdots, \\ m^i_{q,1} \; m^i_{q,2} \cdots m^i_{q,p}, \end{bmatrix}, \tag{6.18}$$

where each column j in $\mathbf{m}_i$ represents the moment vector of all output terminals from the input terminal j and each row k in $\mathbf{m}_i$ represents the moment vector at the output terminal k from all input terminals. Then a moment matrix can be written as

$$M = \begin{bmatrix} \mathbf{m}_0 \; \mathbf{m}_1 \ldots \mathbf{m}_{r-1} \ldots \end{bmatrix}. \tag{6.19}$$

To perform terminal reduction for both inputs and outputs, different moment matrices are constructed. For the output terminal reduction, we define the *output moment matrix* M_O as:

$$M_O = \begin{bmatrix} \mathbf{m}_0^T \\ \mathbf{m}_1^T \\ \vdots \\ \mathbf{m}_{r-1}^T \end{bmatrix}, \tag{6.20}$$

where each column j represents a moment series of output node j from all all inputs' stimuli. Notice that in this way, the output terminals' responses are with respect to all the inputs to make sure they are similar under all the inputs.

Similarly, for input terminal reduction, the *input moment matrix* M_I is defined as:

$$M_I = \begin{bmatrix} \mathbf{m}_0 \\ \mathbf{m}_1 \\ \vdots \\ \mathbf{m}_{r-1} \end{bmatrix}, \tag{6.21}$$

where each column k represents a moment series at all the outputs' nodes from an input node k.

To determine the best order of moments, r, we use the following rule: the number of moments from all the inputs should be equal to or larger than the number of terminals to be merged. The reason is that in the worst case where no terminals can be merged, we should be able to distinguish all the terminals using the moment information. This will become more clear in the following section. As a result, we have

$$rp \geq q \text{ for } M_O, \tag{6.22}$$

$$p \leq rq \text{ for } M_I. \tag{6.23}$$

When $p = q$, then any r will satisfy (6.22) and (6.23). Hence, we can simply select r to be 2 or 3. If we have very large input and output terminals, then we cannot apply SVD to do the terminal reduction directly (this is also true for [31]) as SVD has about $O(n^3)$ complexity, where n is the size of the matrix [44]. However, we can do the terminal reduction in a hierarchical way: we can partition the circuit into k subcircuits such that the terminal in each subcircuit is small enough for SVD, as shown in [34]. More discussions on other special cases will be presented in Subsection 6.6.3.

Moments can be efficiently computed by recursively solving the given circuits using a traditional SPICE-like simulation technique [91]. After we obtain the moment matrix, as shown in (6.20) or (6.21), we proceed to find the optimum number of clusters using singular value decomposition method to be shown in the next section.

6.3 The extended-SVDMOR (ESVDMOR) method

In this section, we present an extended terminal and model order reduction algorithm, ESVDMOR, which improves the SVDMOR method. The basic idea of the new method is to perform the SVD low-rank approximation for the input and output terminals *separately* and use higher-order moment information during the SVD approximation to find true terminal independency and ensure the accuracy of reduced models.

6.3.1 *The ESVDMOR terminal reduction algorithm*

The main idea of the new terminal reduction method is to perform the SVD approximation on the input and output moment response *separately* with the use of high-order moment information. We basically follow the terminal reduction framework of the SVDMOR method [34] but with different moment matrices. But we improve the efficiency of SVDMOR by saving one computation step, as shown later.

The major problem for the SVDMOR method is that both input and output responses are considered at the same time during SVD because of the use of the specific order of block moments $\mathbf{m}_i$. So we cannot accommodate higher order moment information. To mitigate this problem, we create new moment matrices for input and output terminals separately. In this way, we can use higher order moments during the SVD process for input and output responses.

Next, we perform singular value decomposition to both input moment response matrix M_I and output moment response matrix M_O.

$$M_I = U_I \Sigma_I V_I^T \approx U_{I_{k_i}} \Sigma_{k_i} V_{I_{k_i}}^T, \tag{6.24}$$

$$M_O = U_O \Sigma_O V_O^T \approx U_{O_{k_o}} \Sigma_{k_o} V_{O_{k_o}}^T, \tag{6.25}$$

where Σ_{k_i} is a $k_i \times k_i$ diagonal matrix and k_i is the number of significant singular values for matrix M_I; $V_{I_{k_i}}^T$ is a $k_i \times p$ matrix. Similarly, Σ_{k_o} is a $k_o \times k_o$ diagonal matrix and k_o is the number of significant singular values for matrix M_O; $V_{O_{k_o}}^T$ is a $k_o \times q$ matrix.

Then we can perform the low-rank approximation for the input and output position matrix B and C respectively.

$$B = B_r V_{I_{k_i}}^T, \tag{6.26}$$

$$L = V_{O_{k_o}} L_r, \tag{6.27}$$

where $B_r \in R^{n \times k_i}$ and $L_r \in R^{k_o \times n}$ are obtained by computing the Moore–Penrose pseudo-inverses of $V_{I_{k_i}}$ and $V_{O_{k_o}}$ respectively.

$$B_r = BV_{I_{k_i}}(V_{I_{k_i}}^T V_{I_{k_i}})^{-1}, \tag{6.28}$$

$$L_r = (V_{O_{k_o}}^T V_{O_{k_o}})^{-1} V_{O_{k_o}}^T L. \tag{6.29}$$

Notice that both $V_{I_{k_i}}$ and $U_{O_{k_o}}$ are orthonormal matrices, i.e. $V_{I_{k_i}}^T V_{I_{k_i}} = I$ and $U_{O_{k_o}}^T U_{O_{k_o}} = I$. Therefore, (6.28) and (6.29) can be further simplified as

$$B_r = BV_{I_{k_i}}, \tag{6.30}$$

$$L_r = V_{O_{k_o}}^T L. \tag{6.31}$$

Therefore, we save one computation step compared with the SVDMOR method [34].

As a result, the circuit transfer function now becomes

$$H(s) = V_{O_{k_o}} L_r (G + Cs)^{-1} B_r V_{I_{k_i}}^T. \tag{6.32}$$

Consequently we get a terminal reduced subsystem with transfer function $H_r(s)$.

$$H_r(s) = L_r (G + Cs)^{-1} B_r. \tag{6.33}$$

For this subsystem, the standard model order reduction techniques [32,37,85,91,113] can now be applied.

Consider both terminal and model order reductions; we can obtain the order reduced transfer function $\hat{H}_r(s)$,

$$\hat{H}_r(s) = \hat{L}_r (\hat{G} + \hat{C}s)^{-1} \hat{B}_r, \tag{6.34}$$

where

$$\begin{aligned} \hat{G} &= V^T G V; \ \hat{C} = V^T C V; \\ \hat{B}_r &= V^T B V_{I_{k_i}}; \ \hat{L}_r = V_{O_{k_o}}^T L V, \end{aligned} \tag{6.35}$$

where V is the projection matrix for reducing system of (6.33).

The final reduced transfer function becomes

$$\hat{H}(s) = V_{O_{k_o}} \hat{L} (\hat{G} + \hat{C}s)^{-1} \hat{B} V_{I_{k_i}}^T. \tag{6.36}$$

6.3.2 ESVDMOR algorithm flow

In this subsection, we give the whole combined terminal and model order reduction flow of the ESVDMOR method.

Algorithm 6.1: EXTENDED SVDMOR: ESVDMOR

1. Compute the block moments $\mathbf{m}_i$ up to the $(r-1)$th order using (6.4).
2. Construct the input and the output moment response matrices defined in (6.21) and (6.20) respectively.
3. Perform the SVD-based low-rank approximation on the position matrices B and L in (6.1) using (6.30) and (6.31).
4. Perform the normal Krylov-subspace-based MOR on the terminal reduced system (6.33) and perform the transformations using (6.35).
5. Compute the final reduced system $\hat{H}(s)$ using (6.36).

6.3.3 Numerical examples for ESVDMOR method

In this subsection, we compare the SVDMOR and ESVDMOR method on one RC example. We only consider the DC moment for the SVDMOR method.

The example RC circuit, *net27*, has 14 inputs and 118 outputs with model order of 182. Because there are more outputs than inputs, the number of independent input terminals can be determined by using only the DC moment as M_I. However, higher-order moment information is needed to find the number of independent output terminals. Here we use first nine order block moments to construct the output moment matrix M_O, which considers the worst case that all the output terminals are independent to each other.

The singular values for $\mathbf{m_0}$, which is used by SVDMOR, M_I, M_O are shown in Table 6.1. From the table we can see that we only need one input and one output terminal after terminal reduction by using the SVDMOR method. By using M_I and M_O in the ESVDMOR method, we find that there are more than one dominant singular value of M_O. We choose one input and five outputs to make the reduced model more accurate.

Table 6.1 *Singular values of DC moment, input moment matrix and output moment matrix of the circuit net27.*

Index	$\mathbf{m_0}$	M_I	M_O
1	5.1587	5.1587	19.828
2	3.9883×10^{-14}	3.9883×10^{-14}	4.4677
3	1.6681×10^{-14}	1.6681×10^{-14}	1.6517
4	–	–	0.3045
5	–	–	0.0348
6	–	–	2.4611×10^{-3}
7	–	–	1.6134×10^{-4}

To compare the accuracy between the SVDMOR and ESVDMOR methods in a fair way, we make sure that the reduced models for both algorithms have the same number of poles. If we use the same order of block moments, then more terminals will lead to more poles using the Krylov-subspace MOR methods.

For circuit *net27*, six poles are used to approximate the original model for both SVDMOR and ESVDMOR methods. The results are shown in Figure 6.2, which shows the frequency responses corresponding to the second input and the tenth output.

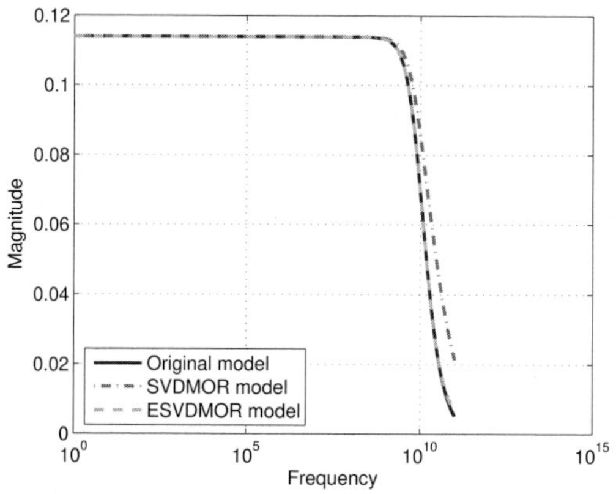

Figure 6.2 Frequency responses from SVDMOR and ESVDMOR for *net27* circuit.

From these frequency response results, we can see that the ESVDMOR model could match the original model up to 5 GHz. In contrast, SVDMOR reduced model can only match frequencies up to 500 MHz. This clearly shows the advantage of the ESVDMOR method over the SVDMOR method when the input and output terminals have different dependency (different ranks in their moment responses matrices).

One may argue that ESVDMOR is more accurate since it uses more terminals. Actually, if we use five inputs and five outputs in SVDMOR method, the results are still not good comparing with our ESVDMOR method under the six poles. The results are shown in Figure 6.3, where *SVDMOR model 2* refers to the SVDMOR results with five input and five output terminals. So simply increasing the number of terminals in SVDMOR does not help to improve the model accuracy.

6.4 Determination of cluster number by SVD

In this section, we present the new *TermMerg* method and present an algorithm to find the best number of independent clusters by using singular value decomposition on the input and output moment matrices discussed in the previous section.

If two terminals have similar timing responses, it means that their moments have

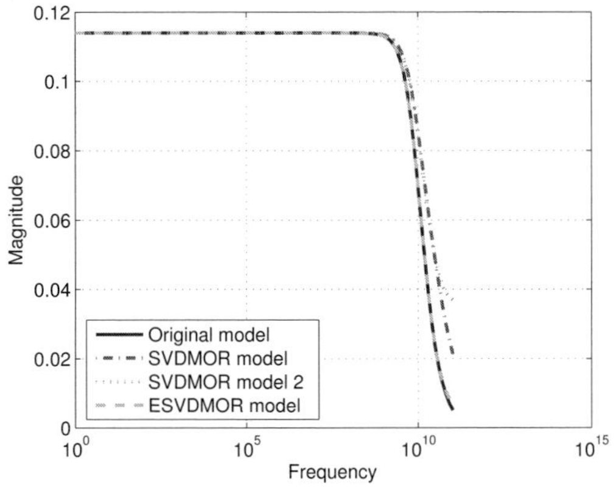

Figure 6.3 Frequency response from SVDMOR and ESVDMOR with different terminals for *net27* circuit.

very similar values. If we have a number of terminals with similar timing behaviors, their moment matrix, where each moment series is a column or a row, will be a low-rank matrix. Singular value decomposition is very efficient to deal with rank-deficient matrices and it can reveal a great deal about the ranks and structure of a matrix, which motivates us to find the optimal number of clusters based on the moment matrices.

For an $m \times n$ matrix A, the SVD decomposition of A is

$$A = U_{m\times m}\Sigma V_{n\times n}^T, \tag{6.37}$$

where $U_{m\times m}$ and $V_{n\times n}$ are orthogonal matrices, $U_{m\times m}^T U_{m\times m} = I$ and $V_{n\times n}^T V_{n\times n} = I$, $\Sigma = \text{diag}(\sigma_1, \sigma_2, \ldots, \sigma_{\min(m,n)})$, σ_i are called singular values and $\sigma_1 \geq \sigma_2 \geq \cdots \geq \sigma_{\min(m,n)}$.

Before we proceed to present the cluster number determination method, we review an important result from the SVD decomposition [44].

For an SVD decomposition of a matrix A, if $k < r = \text{rank}(A)$ and

$$A_k = \sum_{i=1}^{k} \sigma_i u_i v_i^T, \tag{6.38}$$

then

$$\min_{\text{rank}(B)=k} \| A - B \|_2 = \| A - A_k \|_2 = \sigma_{k+1}, \tag{6.39}$$

where $\{u_i\}$ and $\{v_i\}$ are the left and right singular vectors respectively. Equation (6.39) reflects the fact that the rank-k approximation matrix A_k, is just σ_{k+1} away from original matrix A in terms of norm-2 distance.

In summary, for a matrix, A, SVD can do two things. The first thing is that SVD can tell us how many row or column are independent (the true rank of A) in a numerical way. Second, SVD can give a good low-rank approximation to the original matrix $A \approx A_k = \sum_{i=1}^{k} \sigma_i u_i v_i^T$. But it is difficult to find the direct relationship between those left and right singular vectors u_i and v_i with the columns of the matrix A as those vectors are dominant (independent) columns of the original matrix in a different coordinate. But the rank information is still valid for the columns in the original matrix A.

In our problem, we only use the true (low) rank information provided by the SVD instead of the singular vectors. Specifically, we look at the singular values and select the sufficient small singular value σ_{k+1} to select the correct rank information. In the moment matrix, each column represents the moment response for an input or an output terminal. If two terminals have the same response, they should have the same column numerically and thus they are dependent. Hence, the selected rank number k can essentially be viewed as the number of clusters we expect as SVD essentially reveals the true rank of a matrix in a numerical way. Since the rank of matrix A also reflects the number of independent columns in the given matrix A, so it is natural for us to use k as the cluster number. Hence, the true rank information by SVD is a good guide for a later K-means-based clustering method to find the independent terminal set.

We notice that a similar approach was applied to efficient clustering of video images to generate a compact representation of a video sequence to enable fast access to video contents [69]. The method performs the SVD on a transformed matrix, which corresponds the one-to-one with the original matrix for all its column vectors and then performs the K-mean clustering on the transformed matrix to speed up the clustering process. But the clustering essentially is still performed on the columns of the original matrix.

We set a threshold ζ to denote the small singular value σ_{k+1}. Typically ζ is set to 10^{-3} in our experiments, which means we regard σ_{k+1} as the sufficient small singular value only if $\sigma_{k+1} \leq \zeta$.

In the new method, we select the approximation order not only based on the absolute value of the sufficient small singular value, but also on the relative ratio between two adjacent singular values. If the relative ratio between two adjacent singular values is close to 1, that means there is no big difference if the approximation order is increased by one. In this condition, we will keep increasing the approximation order until the ratio becomes small enough. Thus we first define a threshold ε. Then we compare two adjacent singular values σ_{k+1} and σ_k. If $\sigma_{k+1}/\sigma_k \leq \varepsilon$ and $\sigma_{k+1} \leq \zeta$, then k is our choice. Typically ε is set to 10^{-3} in the experiments.

6.5 K-means clustering algorithm

After we determine the number of clusters for the output (or input) terminals, say k, we proceed to group the output (or input) terminals into the k clusters.

In the new approach, we apply a so-called K-means algorithm to determine each cluster [26]. K-means is the widely used clustering algorithm for partitioning (or clustering) N data points into k disjoint subset S_j containing N_j data points to minimize the sum-of-squares criterion:

$$J = \sum_{j=1}^{k} \sum_{i \in S_j} |\mathbf{x}_i - \varphi_j|^2, \tag{6.40}$$

where $\mathbf{x}_i$ is a vector representing the ith data point, and φ_j is a vector, representing the geometric centroid of the data points in S_j. It has wide applications in data mining and data analysis.

However, the K-means clustering algorithm is very sensitive to the initial choice of cluster centers and the number of clusters [26]. Only the appropriate number of clusters produces the best approximation result. Fortunately, in this method, the cluster number k can be first obtained from SVD decomposition on the moment matrix as mentioned above.

For a general moment matrix M ($M = M_O$, or $M = M_I$ for different terminal reductions), it has l terminals to be merged. Let $M = [\mathbf{c}_1, \mathbf{c}_2, \ldots, \mathbf{c}_l]$.

The clustering algorithm is shown in Figure 6.4 which is based on the K-means clustering scheme.

```
K-MEANSCLUSTER
CONSTRUCT_CLUSTER(k)
1  select k seed vectors out of c_1, c_2, ..., c_l as k centroids
2  put the rest of the unselected vectors into the nearest cluster S_j
3  compute φ_j for each cluster S_j
4  do re-cluster all c_i into k clusters according to φ_j of each cluster
5        recompute φ_j for each cluster S_j
6  until no change in φ_j
7  return S_1, S_2, ..., S_k and φ_1, φ_2, ..., φ_k

SELECT_REP_VECTOR(S_1, S_2, ..., S_k)
1  for i ∈ S_j, compute ||d_i||^2 = ||c_i − φ_j||^2
2        find R_j = c_i so that min{||d_i||^2}
3  return R
4  end
```

Figure 6.4 K-means clustering algorithm. Reprinted with permission from [75] (c) 2005 IEEE.

There are two steps (algorithms) in the clustering method. The first step, CONSTRUCT_CLUSTER(k), clusters the given l vectors into k clusters ($l > k$). The second step SELECT_REP_VECTOR($S_1, S_2, \ldots, S_k$)) takes the output of the clustering algorithm as the input and finds the representative vector for each cluster. The representative vectors R_j will be kept during the terminal reduction. All the other unselected vectors will be suppressed.

The basic idea of CONSTRUCT_CLUSTER() is to dynamically find the k clusters so that all the vectors in a cluster have the closest distance to its geometric centroid vector φ_j of cluster S_j. CONSTRUCT_CLUSTER() uses an iterative algorithm that minimizes the sum of distances from each vector to its cluster centroid vector φ_j, over all clusters. This algorithm moves vectors among clusters until the sum cannot be decreased any more.

The second algorithm SELECT_REP_VECTOR() finds the representative vector for each given cluster S_j. This has been achieved by calculating the distance between the centroid and each vector belonging to this cluster to find the terminal with the closest distance to the centroid.

Since we use the representative terminal to represent all the other terminals in a cluster, the transfer functions of the reduced one (thus their zero) will be different from their original ones. But we ascertain that their expanded moment form will be as close as possible among terminals in a cluster (as we work on the moment matrix). Hence we expect the zeros of the two transfer functions (for the representative one and the reduced one) will be similar.

6.6 TermMerg algorithm

In this section, we first present the whole terminal reduction flow of the TERMMERG algorithm. Then we give the compact modeling flow, which combines the terminal reduction with traditional model order reduction. Finally we discuss some practical considerations of the algorithm.

6.6.1 *TermMerg algorithm flow*

In this subsection, we present the flow of the TermMerg method.

Algorithm 6.2: TERMMERG

1. Construct the input and the output moment matrices.
2. Perform the SVD on the input and/or the output moment matrices to find the best cluster numbers.
3. Invoke K-MEANSCLUSTER to find the all the representative terminals for each cluster.

After the terminal reduction, the input position matrix B and the output position matrix C in (6.15) will be modified to include only the representative terminals. We can define terminal selection matrices for this operation on B and C.

6.6.2 Modeling flow based on combined terminal and model order reduction

After the terminal reduction, we will apply the traditional model order reduction technique on the terminal-reduced circuit, as shown in Figure 6.5. The new terminal reduction can work with any existing model order reduction techniques, such as projection-based methods [85], TBR-ased methods [87],or hierarchical reduction methods [121].

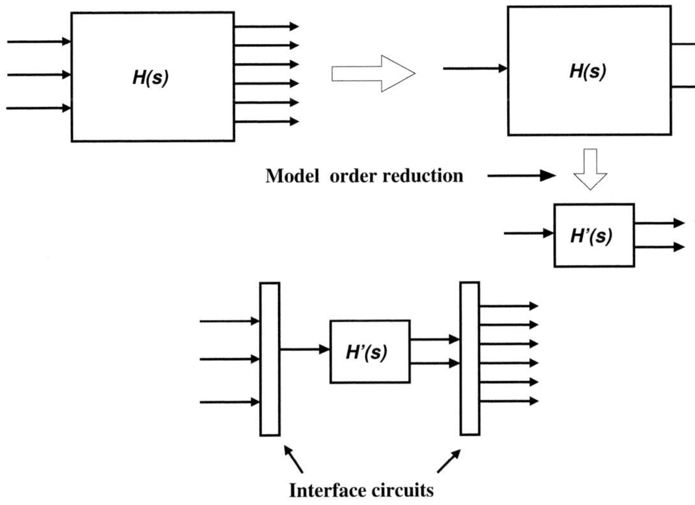

Figure 6.5 The reduction flow of combined terminal and model order reductions.

To use the terminal and order reduced models, we may need to build simple interface circuits to connect the original terminals to the representative terminals. For output terminals, we may just physically connect the representative terminals to all the reduced terminals they represent (no interface circuitry is required). For input terminals, to avoid adverse coupling among different sources with low impedance, we may use controlled sources (for instance, urrent-controlled current sources (CCCS)) to connect several input sources together. For input sources with high impedance, we can just physically connect them together (again, no interface circuitry is required). To use the terminal and order reduced models, we may need to build simple interface circuits to connect the original terminals to the representative terminals. For output terminals, we may just physically connect the representative terminals to all the fan-out nodes that the reduced terminals connect before (i.e., no interface circuitry is required). Figure 6.6 illustrates a simple interface circuit for the input and output terminals. The admittance for each CCCS is the corresponding input admittance of the represented input terminal, which can be passively realized using Foster's realization method [93].

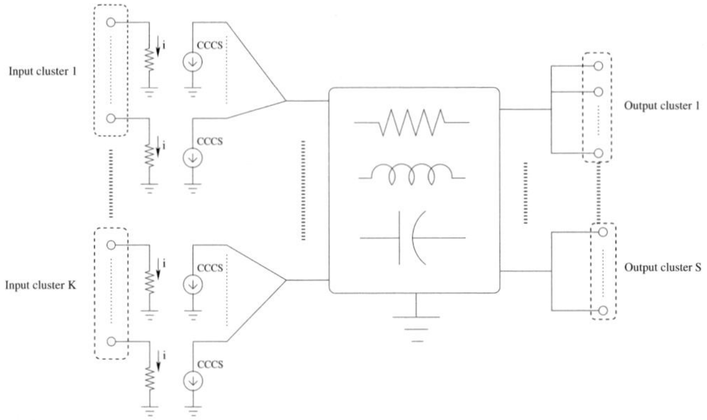

Figure 6.6 Simple interface circuit.

6.6.3 Practical implementation and consideration

We have mentioned that we determine the order of moments r based on the fact that the number of moments in the moment series from all the inputs should be equal to or larger than the number of terminals to be merged, as shown in (6.22) and (6.23). In this way, we can obtain the complete information of the moment matrix after singular value decomposition. However, if we have many outputs and a few inputs, we end up with large r. In other words, we have to use very high order moments. But high order moments are actually not very informative as they contain mainly the dominant pole information numerically [91]. This is also the case for interconnect circuits with many inputs and a few outputs. On the other hand, a large number of outputs or inputs does not imply that the circuit has more independent outputs and inputs, or clusters. Practically, the final cluster number may still be very small compared with the number of outputs and inputs. In this case, we do not need many high order moment information to distinguish those terminals. This motivates the following schemes to solve this problem.

We pre-define an effective cluster number q_e, which is smaller than the number of terminals q to be merged but is typically larger than the resulting number of clusters. For example, we may define $q_e = \lceil q/i \rceil, i = 2, 3, \ldots$ depending on circuits, where the function $\lceil x \rceil$ means rounding the x to the nearest larger integer. Then the order of moments r would be equal to or larger than q_e/p. If the cluster's number is equal to q_e after SVD method at some threshold, that means the pre-defined number of clusters is too small to find the optimal cluster number at this threshold. We then either increase the threshold value (at cost of more approximation errors) or increase the pre-defined clusters' number q_e to re-cluster the terminals. If we already have had some prior-knowledge about the circuit terminals, it will be very helpful to choose the effective cluster number q_e.

Another way is to use a hierarchical terminal reduction scheme, which is suitable for a large number of resulting clusters. Specifically, we first partition the merging terminals into a number of groups so that their corresponding moment matrix M

will not need to use the higher moments. After this we perform SVD on each group and find several representative terminals from each group using K-means clustering method. After this we perform SVD on all the representative terminals to find the best global cluster number. Then we perform the K-means-based clustering algorithm on the top of those selected representative terminals to find the global representative terminals. We may have several hierarchical levels when the number of terminals to be merged is large.

Another issue is that there may exist DC paths from the inputs to the outputs. In this case, (6.16) will be written as

$$H(s) = C(sI - A)^{-1}B + D,$$

and the zeroth moment will become $\mathbf{m}_0 = -CA^{-1}B + D$. If the zeroth moments for those DC-coupled terminals are quite different, they will probably not be clustered together. So the method can still be applied.

6.6.4 *Passivity analysis*

Passivity is an important issue for compact modeling. First we show that the proposed TermMerg terminal reduction algorithm can always guarantee the passivity of reduced models when the model order reduction is passive, regardless of the input and output position matrices B and L. We have the following result:

Proposition 6.1. The TermMerg method can always lead to passive models for any RLCK circuits.

The proof is obvious. For any given RLCK circuit with sets of input and output terminals, after the TermMerg reduction, we have a new set of input and output terminals. To make the terminal-reduced system passive, one can use two reduction approaches. One is by means of the projection-based method. In this case, one first makes all the remaining terminals bidirectional ones (they are both inputs and outputs at the same time). As a result, we have the same terminal reduced input and output position matrices. In this case, the projection-based method can generate the passive model for this circuit.

The second method is by means of passive truncated balanced realization (TBR), which can make passive reduction of RLCK circuits with different B and L^T directly but at higher computation costs [88].

As we can see, as long as the model order reduction is a passive process, the combined terminal and model order reduction process is passive for the new method.

However, this is not the case for SVDMOR as SVDMOR is strongly coupled with the model order reduction process. We analyze two cases. First we have $B \neq L^T$. In this case, B_r and L_r^T are not identical and V_k and V_k are not identical either. As a result, the projection-based MOR method cannot produce passive models. This is also true for passive TBR method as V_k and V_k are not identical.

One may think that one can make all the terminals as bidirectional ones by making B and L^T equal. But we show below that such a simple strategy for enforcing passivity may significantly reduce the terminal reduction qualities.

Table 6.2 lists the singular value results after the SVD on the 0th and 1st moment matrices for circuit *net1026* with both input and output terminals treated as bidirectional ones.

Table 6.2 *Singular values of the DC admittance moment, 1st order admittance moment matrices of the circuit net1026 when all the terminals are treated as bidirectional.*

Index	$\mathbf{m_0}$	$\mathbf{m_1}$
1	0.58970	0.42268
2	0.55093	0.33515
3	0.50215	0.30440
4	0.41875	0.28897
5	0.31229	0.25942
6	0.064575	0.24131
⋮	⋮	⋮
258	0.0023181	0.0099492
259	0.0013246	0.0099117
260	0.00059548	0.0098789
261	0.0001499	0.0098567
262	2.9535×10^{-17}	0.0096795

From Table 6.2, we can see that the singular values decay very slowly and the terminals cannot be reduced very much. Actually the numerical rank of the moment matrix $\mathbf{m_i}$ (given by Matlab rank command) is 261 for $\mathbf{m_0}$ and 262 for $\mathbf{m_1}$, which is the full rank of the moment matrix. This is true for other test cases. As a result, we can observe that SVDMOR is not efficient for reducing bidirectional terminals. One obvious reason is that many output terminals now become input terminals. So the responses excited by those terminals have to be considered for all the other terminals, which make the reduction more difficult or almost impossible.

In the second case, we consider $B = L^T$. To ensure passivity, we require that B_r are L_r^T are the same and V_k and V_k are identical as well owing to the requirements of the congruence transformation. As a result, the moments $\mathbf{m_i}$ must be symmetric. It can be proved that when the inputs to the original models consist only of current sources or voltage sources for RLCK circuits, $\mathbf{m_i}$ is symmetric. If both current and voltage sources are present, $\mathbf{m_i}$ will not be symmetric and terminal reduction by SVDMOR will not ensure the passivity. But as we have already shown, $B = L^T$ will lead to a reduction of bidirectional terminals, which cannot be reduced effectively by SVDMOR.

In short, we observe that SVDMOR in general not lead to passivity terminal reduction. When it does produce passive terminal reduction, it does not work well (or does not work at all). In contrast, TermMerg can always ensure passive terminal reduction by using a passive model order reduction method.

6.7 Numerical examples

The new method has been implemented in MATLAB. We tested the terminal merging algorithm on a number of real interconnect circuits from industry.

The first interconnect circuit *net1026* is a one-bit line circuit from a SRAM circuit in 180 nm technology. This network contains 525 resistors, 772 capacitors, 6 drivers, and 256 receivers. We perform the K-means-based *TermMerg* algorithm to reduce both receiver (output) terminals and driver (input) terminals. Since there are many outputs (256 receivers) compared with a few inputs (6 drivers), in this case, we set the order of moments in the input moment matrix to 1 and the order of the moments in the output moment matrix is computed as $256/6 = 42$. But in this case we set the effective cluster number $q_e = 30$. The output moment order is $r = q_e/6 = 5$.

By using singular value decomposition, the optimal number of clusters is found to be 5 if we define the threshold $\varepsilon = 10^{-3}$. The dominant singular values for the M_I, M_O are listed in Table 6.3.

6.7.1 Comparison with the SVDMOR method

One problem with the SVDMOR method is that only a specific moment is used to determine the correlations of both input and output terminals. The specific moment, like the DC moment, can give a good estimation of terminal correlations on the low frequencies, but they may not be accurate at the high frequencies. This is specially true for circuits having small numbers of inputs and large numbers of outputs or vice versa.

Figure 6.7 shows the SVDMOR reduction results for the interconnect circuit, *net1026*, in frequency domain. We perform first the terminal reduction and then Krylov-subspace-based MOR. We keep up to the 2nd order of block moments in the MOR for both the SVDMOR and TermMerg methods. SVDMOR based on $\mathbf{m}_0$ reduces the terminals to only one input and one output based on the singular values, as shown in Table 6.3. For TermMerg, we have one input and five outputs after terminal reduction.

From Figure 6.7 we can see that the result from SVDMOR does not match well with the original circuit at high frequencies while the proposed method has a better matching at the same high frequencies.

Accuracy loss at high frequency after terminal reduction reflects the fact that the only specific order of moments (DC moments) are used during the SVD-based terminal reduction and DC moments will lead to less accurate terminal relationship at the high frequency ranges. This problem can easily be seen when the numbers of

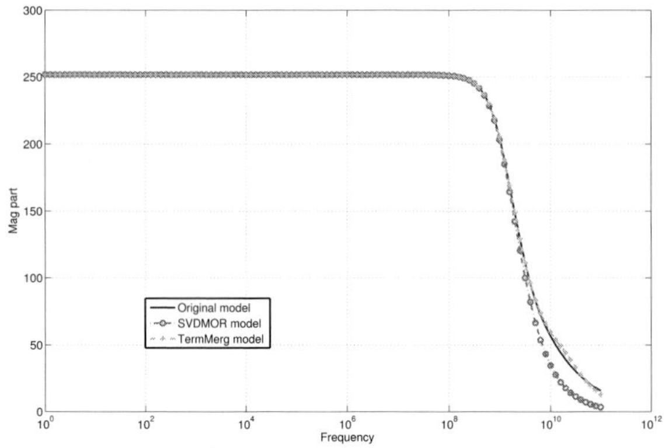

Figure 6.7 Frequency impedance responses from the SVDMOR method for *net1026* circuit.

Table 6.3 *Singular values of DC admittance moment, input moment matrix and output moment matrix of the circuit net1026.*

♯	$\mathbf{m_0}$	M_I	M_O
1	5789.5	5789.5	46355
2	1.4364×10^{-12}	1.4364×10^{-12}	666.95
3	3.7177×10^{-13}	3.7177×10^{-13}	22.558
4	–	–	0.32084
5	–	–	0.00266
6	–	–	1.4549×10^{-5}
7	–	–	5.6469×10^{-8}

input and output terminals are quite different as in case of *net1026*. For this circuit, we can compare 256 (DC) response at 256 output terminals for each input terminal, so we have a better picture for determining their correlations even using the DC moments. While for each output port, we can only compare the responses from six inputs, the terminal relationship will not be accurate with just DC moments. Once we use high-order moments, we obtain better correlations of the output terminals and, thus, the reduced model is more accurate at the high frequencies, as shown in Table 6.3 and Figure 6.7.

6.7.2 *Clustering results*

The final clustering results from TERMMERG are shown in Table 6.4. The first column is the number of the cluster series. The second column is the representative terminal of each cluster. All the terminals in each cluster are placed in the third column.

Table 6.4 *Output clustering results for the one-bit lines circuit net1026. Reprinted with permission from [75] (c) 2005 IEEE.*

Cluster Index	Representive terminal	Clustered terminals
1	Rcv206	Rcv151, Rcv152, ..., Rcv256
2	Rcv58	Rcv39, Rcv40, ..., Rcv77
3	Rcv19	Rcv1, Rcv2, ..., Rcv38
4	Rcv98	Rcv78, Rcv79, ..., Rcv119
5	Rcv144	Rcv120, Rcv121, ..., Rcv150

Figure 6.8 plots the distribution of terminals (x-axis) with respect to the cluster index number (y-axis).

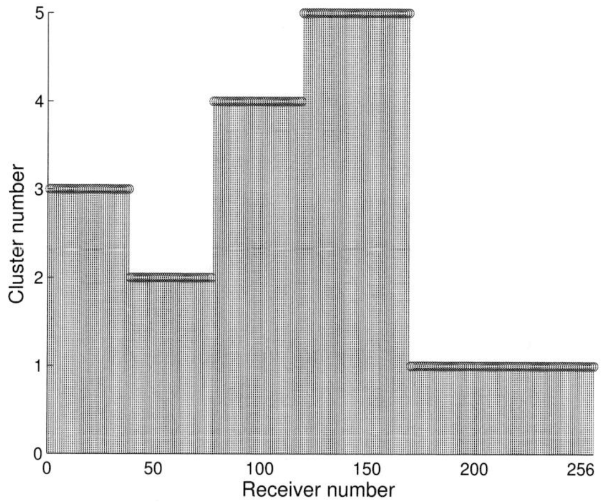

Figure 6.8 Output terminal distribution for each cluster for *net1026* circuit. Reprinted with permission from [75] (c) 2005 IEEE.

Then we go back to the time domain to validate the effectiveness of the method. We add a voltage source to a driver input to view the step responses at other receivers. Figure 6.9 shows the responses of five representative terminals. If we compare the 50% delay time, the delay time difference among them is approximately 10–20 ps, which is quite different. The enlarged local waveforms in Figure 6.9 are shown in Figure 6.10.

If we plot more responses for all the suppressed terminals in one cluster, for instance *receiver97* and *receiver99*, whose representative terminal is *receiver98*, we cannot tell the difference in responses between these reduced terminals and their representative terminal, *receiver98*, for the delay time, as shown in Figure 6.11. Detailed analysis shows that the delay time differences among these terminals are only about 1–2 ps, which is clearly shown in Figure 6.12. In other words, if we allow

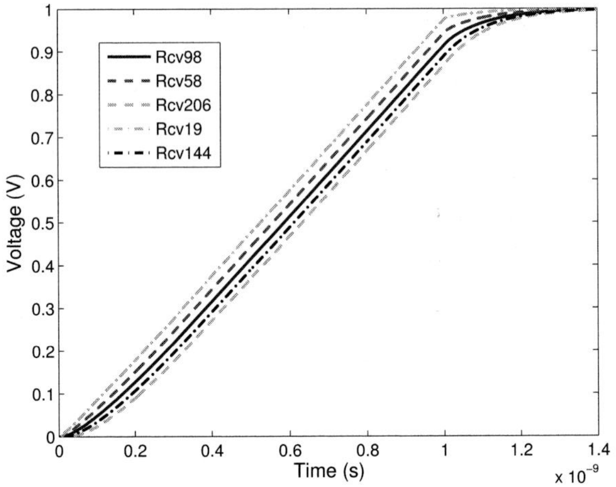

Figure 6.9 Step responses of representative output terminals. Reprinted with permission from [75] (c) 2005 IEEE.

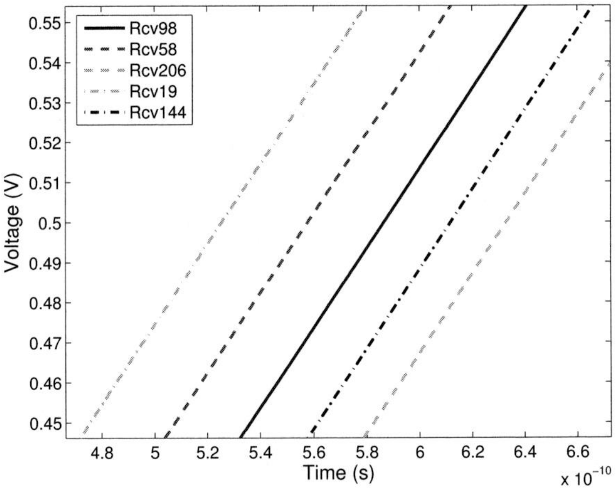

Figure 6.10 Comparison of 50% delay time among representative output terminals. Reprinted with permission from [75] (c) 2005 IEEE.

1–2 ps delay variations, those terminals can be viewed as the same terminal.

At this point, we can say that it is reasonable to use the response at *receiver98* to represent the responses at the suppressed terminals in its cluster, such as *receiver97* and *receiver99*. Considering the process variations and other environmental variations, it is possible that we can combine them into one terminal. Also we can improve the accuracy of this method by relaxing the threshold ε to generate more clusters.

For the input terminal merging, we need to cluster the six input (six drivers)

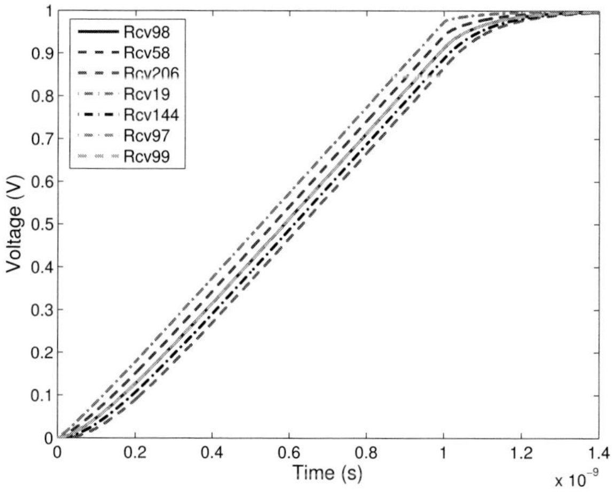

Figure 6.11 Step responses of representative output terminals and two suppressed outputs. Reprinted with permission from [75] (c) 2005 IEEE.

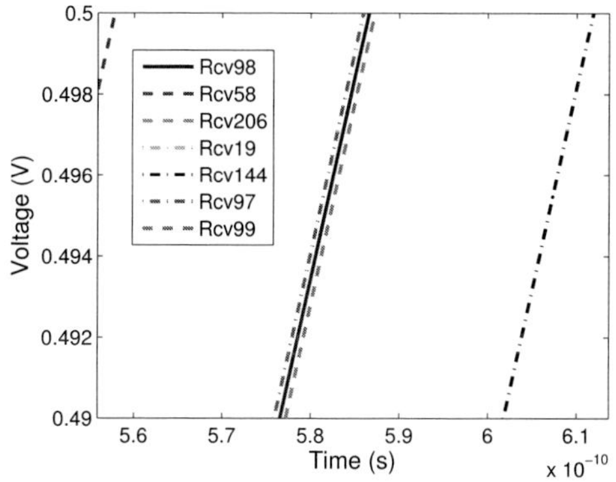

Figure 6.12 Comparison of 50% delay time among representative output terminals and two suppressed outputs. Reprinted with permission from [75] (c) 2005 IEEE.

terminals. By using the same threshold level $\varepsilon = 10^{-3}$, the cluster number is 5. Only the terminals at *driver3* and *driver4* could be merged together.

The second example, *net27*, is a clock-tree circuit, also in 180 nm technology. It contains 167 resistors, 654 capacitors, 14 drivers, and 118 receivers. For the output reduction, we set the effective cluster number $q_e = 118/2 = 59$. Then we only need $r = q_e/14 \approx 4$ orders of moments to format the output matrix M_O. After the SVD step, as shown in Table 6.1, it is obvious to select the cluster number as $k = 5$ at the given $\varepsilon = 1 \times 10^{-2}$.

We also present its distribution of terminals for different clusters in Figure 6.13 when we select the cluster number $k = 5$. The representative terminals are *receiver98*, *receiver18*, *receiver110*, *receive36*, and *receiver84* corresponding to clusters

1–5.

Notice that the receiver numbers were not assigned based on closeness between terminals for *net27*. But for circuit *net1026*, it seems that receivers are numbered based on the closeness between them.

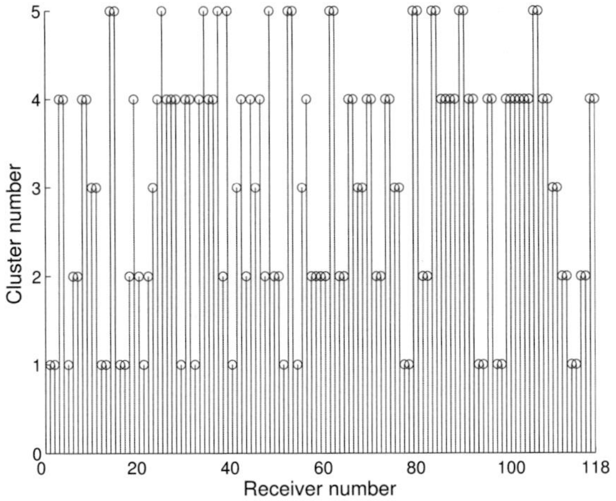

Figure 6.13 Output terminal distribution for each cluster for *net27* circuit. Reprinted with permission from [75] (c) 2005 IEEE.

The third example, *net38*, is a clock-tree circuit also. It includes 98 resistors, 342 capacitors, 18 drivers, and 48 receivers. We use $r = (48/18) \approx 3$ moments to format the output matrix M_O as well as input matrix M_I.

After the SVD, the input moment matrix M_I has the following singular values: $\Sigma = \text{diag}(1.03 \times 10^5, 42.88, 7.35, 2.05, 0.83, 0.17, 3.74 \times 10^{-5}, \ldots)$. If we set the threshold $\varepsilon = 1 \times 10^{-3}$, the cluster number is 16, which is almost equal to the number of inputs. However we notice that there is a big magnitude drop between two singular values 0.17 and 3.74×10^{-5}. If we relax the threshold to $\varepsilon = 3 \times 10^{-3}$, the number of clusters is six. The distribution of terminals for different clusters is presented in Figure 6.14.

For the output terminal reduction, the number of clusters will be 48 if the threshold is set to $\varepsilon = 3 \times 10^{-3}$, which is the number of the output terminals. That means we are not able to cluster the output terminals at this threshold level. After we relax the threshold to $\varepsilon = 6 \times 10^{-2}$, we can get the cluster number that is 13. We show the output assignment results in Figure 6.15.

6.8 Summary

In this chapter, we presented a novel method, named *TermMerg*, to efficiently reduce the terminal number of general linear interconnect circuits considering delay

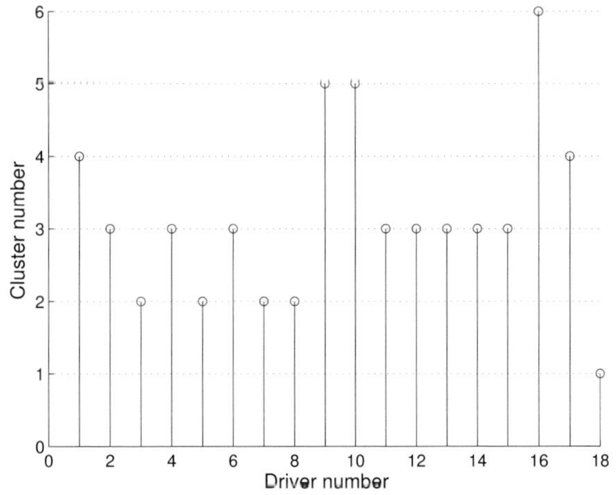

Figure 6.14 Input terminal distribution for each cluster for circuit *net38*.

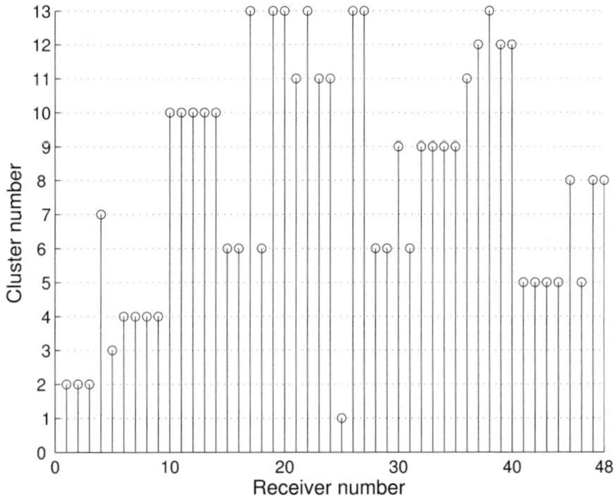

Figure 6.15 Output terminal distribution for each cluster for circuit *net38*.

uncertainty. Terminal reduction can lead to more compact order-reduced models. The new method is based on the high-order moment responses of terminals in frequency domain as the metrics for the timing or delay. We first applied a singular value decomposition method to determine the best number of clusters based on the low-rank approximation. After this, the K-means-based clustering algorithm was used to cluster the moments of the terminals into the different clusters. An improved SVDMOR, *ESVDMOR*, has been introduced, which improves the SVDMOR in terms of efficiency and accuracy. Experimental results on a number of real industry interconnect circuits demonstrated the effectiveness of the new method.

7 Vector-potential equivalent circuit for inductance modeling

As VLSI technology advances with decreasing feature size as well as increasing operating frequency, inductive effects of on-chip interconnects become increasingly significant in terms of delay variations, degradation of signal integrity, and aggravation of signal crosstalk [72, 116]. Since inductance is defined with respect to the closed current loop, the loop-inductance extraction needs to specify simultaneously both the signal-net current and its returned current. To avoid the difficulty of determining the path of the returned current, the partial element equivalent circuit (PEEC) model [101] can be used, where each conductor forms a virtual loop with infinity and the *partial inductance* is extracted.

To model inductive interconnects accurately in the high frequency region, RLCM (M here stands for mutual inductance) networks under the PEEC formulation are generated from discretized conductors by volume decomposition according to the skin-depth and longitudinal segmentation according to the wavelength at the maximum operating frequency. The extraction based on this approach [59, 83, 84] has high accuracy but typically results in a huge RLCM network with densely coupled partial inductance matrix L. A dense inductively coupled network sacrifices the sparsity of the circuit matrix and slows down the circuit simulation or makes the simulation infeasible. Because the primary complexity is a result of the dense inductive coupling, efficient yet accurate inductance sparsification becomes a need for extraction and simulation of inductive interconnects in the high-speed circuit design.

Because the partial inductance matrix in the PEEC model is not diagonally dominant, simply truncating off-diagonal elements leads to negative eigenvalues and the truncated matrix loses passivity [52]. There are several inductance sparsification methods proposed with the guaranteed passivity. The return-limited inductance model [109] assumes that the current for a signal wire returns from its nearest power and ground (P/G) wires. This model loses accuracy when the P/G grid is sparsely distributed. The shift-truncation model [65] calculates a sparse inductance matrix by assuming that the current returns from a shell with shell radius r_0. But it is difficult to determine the shell radius to obtain the desired accuracy. Because the inverse of the inductance matrix, called the K-element (susceptance) matrix is strictly diagonally dominant, off-diagonal elements can be truncated without affecting the passivity [9, 25]. Because K is a new circuit element not included

a_i	ith filament
A^k	kth component of vector potential A
A_i^k	kth component of averaged A at a_i
$\mathcal{B}_{ij}^k$	kth component of the flux from a_i to a_j
$\mathcal{B}_{i0}^k$	kth component of the flux from a_i to vector potential ground
I_i^k	kth component of electrical branch current at a_i
$\hat{I}_i^k$	kth component of magnetic branch current at a_i
$\hat{R}_{ij}^k$	kth component of coupling effective resistance between a_i and a_j
$\hat{R}_{i0}^k$	kth component of ground effective resistance of a_i
V_i^k	kth component of electrical branch voltage at a_i
$\hat{V}_i^k$	kth component of magnetic branch voltage at a_i

Table 7.1 *Table of notations. Reprinted with permission from [135] (c) 2005 IEEE.*

in conventional circuit simulators such as SPICE, new circuit analysis tools considering K have been developed [17, 58]. Alternatively, the double-inversion-based approaches have been proposed in [9, 140]. Using the control volume to extract adjacently coupled effective resistances to model inductive effects, the vector-potential equivalent circuit (VPEC) model was recently introduced [86]. Its sparsified and SPICE-compatible circuit model is obtained based on a locality assumption that the coupling under the VPEC model exists only between adjacent wire filaments.

This chapter presents a cost-efficient and probably passive inductance model: the vector-potential equivalent circuit (VPEC). Because the extraction in [86] requires to optimize the size of the control volume for each filament during integration, it becomes impractical for large-sized interconnects. Moreover, it need locality assumption that may lead to an inaccurate extraction [134]. In contrast, we rigorously derive an accurate VPEC model considering the coupling between any pair of filaments by inverting the partial inductance matrix. We further prove that the resulting circuit matrix for the full VPEC model is passive and strictly diagonal dominant. The diagonal dominance enables truncating small-valued off-diagonal elements to obtain a sparsified VPEC model named the truncated-VPEC (tVPEC) model with guaranteed passivity.

7.1 Vector-potential equivalent circuit

As in FastHenry [59] with the magneto-quasi-static assumption, a conductor can be divided into a number of rectilinear *filaments*. The current density is *constant* over the cross-section of the filament. In this chapter, we use superscripts x, y, and z to denote spatial components of a vector variable. Let $\mathbf{A}$ be the *vector potential*, determined by the distribution of the current density J. Then J^k and A^k are the components in the k-direction ($k = x, y, z$). We further use the subscript i for variables associated with filament a_i ($i \in N$), and every filament a_i has a length l by adequate discretizations in the k-direction. Table 7.1 summarizes the notations used in this chapter.

To model the inductive effect, we start with differential Maxwell equations in terms of **A** [57]:

$$\nabla^2 A^k = -\mu J^k \tag{7.1}$$

$$\frac{\partial A^k}{\partial t} = -E^k, \tag{7.2}$$

where the vector potential **A** and the current density **J** are in the z-direction, **E** is the electrical field, and μ is the permeability constant. Note that the resistive voltage drop by $(-\nabla^k \phi)$ is not included in (7.2) since we are interested in the inductive voltage drop here. Given the distribution of the current density J^k, the vector potential A^k is determined by

$$A^k = \frac{\mu}{4\pi} \int \frac{J^k}{|\mathbf{r} - \mathbf{r}'_i|} \mathrm{d}\tau(\mathbf{r}'_i).$$

To construct the system equation in the form of the integral equation, we apply the volume and the line integration to (7.1) and (7.2) respectively. For filament a_i, (7.1) is integrated within the volume τ_i of filament a_i, using *Gauss' law*:

$$\int_S \mathbf{a} \cdot \mathrm{d}\mathbf{S} = \int_\tau \nabla \cdot \mathbf{a} \mathrm{d}\tau,$$

we can obtain

$$\begin{aligned} -\mu \int_{\tau_i} J^k \mathrm{d}\tau &= \int_{S_i} \nabla A^k \cdot \mathrm{d}\mathbf{S} \\ &= \mathcal{B}^k_{i0} + \sum_{j \neq i} \mathcal{B}^k_{ij}. \end{aligned} \tag{7.3}$$

Note that the surface integral $\int_{S_i} \mathrm{d}\mathbf{S} \cdot \nabla A^k$ is actually the *flux of the gradient of the kth component of the vector potential* caused by the filament current of a_i in τ_i. It consists of the following parts [86]: (i) the flux to infinity (vector potential ground) $\mathcal{B}^k_{i0}$

$$\mathcal{B}^k_{i0} = \int_{S_{i0}} \nabla A^k \cdot \mathrm{d}\mathbf{S}, \tag{7.4}$$

and (ii) the flux to all other filaments a_j $(j \in N, j \neq i)$ $\mathcal{B}^k_{ij}$

$$\mathcal{B}^k_{ij} = \int_{S_{ij}} \nabla A^k \cdot \mathrm{d}\mathbf{S}. \tag{7.5}$$

However, explicitly determining the value of $\mathcal{B}^k_{ij}$ is difficult because it is hard to partition the flux between filament a_i, all other filaments a_j, and the vector potential ground.

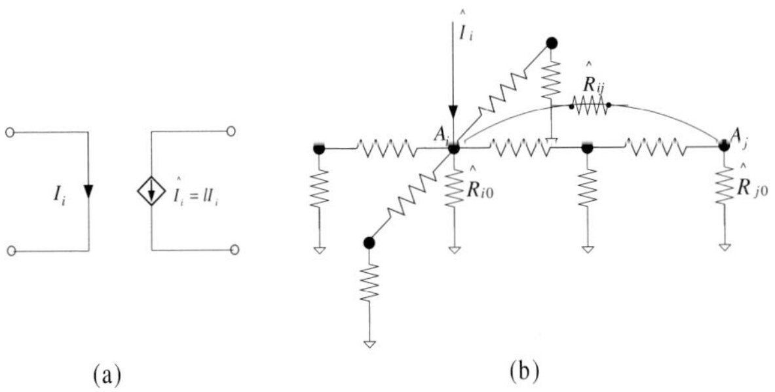

Figure 7.1 (a) Electronic current-controlled vector-potential current source; (b) The Kirchoff current law for vector potential circuit. An invoking vector potential current source is employed at a_i, and the responding vector potential at a_j is A_j^k, determined by the full effective resistance network. Reprinted with permission from [135] (c) 2005 IEEE.

Moreover, integrating (7.2) along the projected length in k-direction of filament a_i leads to:

$$\int_{l_i} \frac{\partial A^k}{\partial t} \mathrm{d}l = - \int_{l_i} E^k \cdot \mathrm{d}l. \tag{7.6}$$

Based on (7.3) and (7.6), we can further construct the circuit-level system equation in the matrix form. By defining the *filament vector potential* [86] as the *average* volume integral of A^k within τ_i (surrounded by the surface S_i):

$$A_i^k = \frac{1}{\tau_i} \int_{\tau_i} A^k(\mathbf{r}) \mathrm{d}\tau(\mathbf{r}). \tag{7.7}$$

We can define an effective coupling resistance

$$\hat{R}_{ij}^k = -\mu \frac{(A_i^k - A_j^k)}{\mathcal{B}_{ij}^k} \tag{7.8}$$

to model (i.e., replace) the mutual inductive coupling between a_i and a_j. In addition, there also exists an effective ground resistance to model the self inductive effect:

$$\hat{R}_{i0}^k = -\mu \frac{A_i^k}{\cdot} \mathcal{B}_{i0}^k \tag{7.9}$$

Because the filament current is invariant along the k-direction, the volume integral of the current density inside the volume τ_i is reduced to lI_i^k, where I_i^k is the electrical current at the cross-section of a_i. Therefore (7.3) becomes the *Kirchoff current law* (KCL) under the full VPEC model:

$$\frac{A_i^k}{\hat{R}_{i0}^k} + \sum_{j \neq i} \frac{(A_i^k - A_j^k)}{\hat{R}_{ij}^k} = lI_i^k, \tag{7.10}$$

where a vector potential current source $\hat{I}_i^k$ can be defined:

$$\hat{I}_i^k = lI_i^k, \tag{7.11}$$

which is controlled by the electrical current I_i^k. An equivalent circuit to illustrate the VPEC KCL equation (7.10) is shown in Figure 7.1. Clearly, we can see the physical meaning of the effective resistance: *given a unit current change at the ith filament, the vector potential observed at jth filament is exactly $\hat{R}_{ij}^k$ when all other filaments are connected to the vector potential ground.*

Similarly for (7.6), we have the following *inductive* nodal voltage equation:

$$l\frac{\partial A_i^k}{\partial t} = V_i^k, \tag{7.12}$$

which describes the relation between the vector potential and its corresponding electrical voltage drop caused by the inductive effect. As a result, a voltage-controlled vector potential voltage source $\hat{V}_i^k$ is:

$$\hat{V}_i^k = V_i^k/l. \tag{7.13}$$

To extract the vector-potential equivalent circuit, the integration-based approach in [86] needs to determine the *localized flux* $\mathcal{B}_{ij}^k$ $(j \in n_i)$, where n_i is the set of filaments adjacent to a_i. The explicit calculation of $\mathcal{B}_{ij}^k$ is hard, and only considering the localized flux, as in [86], loses the accuracy. The integration-based VPEC model in [86] needs to use a control volume for each filament, but no method was presented in [86] to find an accurate control volume. To avoid calculating $\mathcal{B}_{ij}^k$ explicitly and use the locality assumption, this chapter derives a full VPEC model and then present an inversion-based extraction below. The detailed derivation of the full VPEC model is presented in [135], where we obtain following two KCL and KVL equations:

$$\frac{A_i^k}{\hat{R}_{i0}^k} + \sum_{j \neq i, i, j \in N} \frac{(A_i^k - A_j^k)}{\hat{R}_{ij}^k} = \hat{I}_i^k, \tag{7.14}$$

$$\frac{\partial A_i^k}{\partial t} = \hat{V}_i^k, \tag{7.15}$$

where the vector potential current and voltage are related to the electrical branch current and voltage by

$$\hat{I}_i^k = lI_i^k, \qquad \hat{V}_i^k = V_i^k/l. \tag{7.16}$$

Clearly, we can see the physical meaning of the effective resistance by (7.14): *given a unit current change at the ith filament, the vector potential observed at jth filament is exactly $\hat{R}_{ij}^k$ when all other filaments are connected to the vector potential ground.*

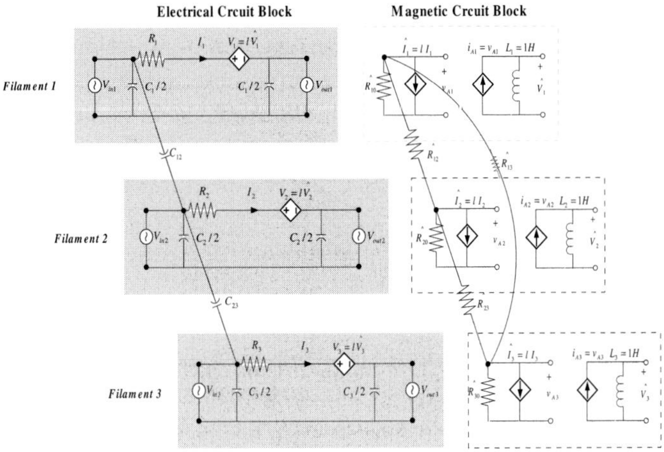

Figure 7.2 Vector potential equivalent circuit model for three filaments. Reprinted with permission from [135] (c) 2005 IEEE.

Furthermore, (7.15) describes the relation between the vector potential and its corresponding electrical voltage drop caused by the inductive effect.

We present a SPICE-compatible VPEC model for three filaments in Figure 7.2. The model consists of two blocks: the electrical circuit (PEEC resistance and capacitance) and the magnetic circuit (VPEC effective resistance and unit inductance). They are connected by the controlled sources. It includes the following components:

1. The resistance R_i and capacitance C_i in the electrical circuit are the same as those in the PEEC model;
2. A dummy voltage source to sense current I_i^k in the electrical circuit controls $\hat{I}_i^k$ in the magnetic circuit;
3. A voltage-controlled current source is used to relate $\hat{V}_i^k$ and $\hat{I}_i^k$ with gain $g = 1$ in the magnetic circuit;
4. A voltage source V_i^k in the electrical circuit is controlled by $\hat{V}_i^k$ in the magnetic circuit;
5. Effective resistances including ground $\hat{R}_{i0}^k$ and coupling $\hat{R}_{ij}^k$ are used to represent the strength of inductances in the magnetic circuit;
6. A unit inductance L_i in the magnetic circuit is to: (i) take into account time derivative of A_i^k; and (ii) preserve the magnetic energy from electrical circuit.

Although the number of magnetic circuit blocks increases with more filaments, sparsified VPEC models will be introduced in Subsection 7.3.2 to greatly reduce coupling resistances in magnetic circuit blocks with preserved passivity. Moreover, because the VPEC model largely reduces reactive elements (i.e., inductance) and its effective resistance is less densely stamped in the MNA (modified nodal analysis) matrix compared with the partial inductance under the PEEC model, the full VPEC model reduces the simulation time compared with the PEEC model.

Note that the summation in KCL (7.14) for the full VPEC model is carried out over each pair of filaments. In contrast, this summation in [86] is carried out only for adjacent filaments. The author of [86] obtained the localized model by modeling the flux $\mathcal{B}_{ij}^k$ as a *current* flow through $\hat{R}_{ij}^k$. It is based on the analogy with the conducting current flow at a surface S (*Ohm's law*):

$$\mathbf{I} = -\sigma \int_S \nabla\phi \cdot \mathrm{d}\mathbf{S} \tag{7.17}$$

Equation (7.17) means that the conducting current $\mathbf{I}(x, y, z)$ is locally related to the flux of the electrical field $\mathbf{E}(x, y, z)$ $(-\nabla\phi)$ on the surface S. Equation (7.17) is correct because electrons only locally transport in the conductor. However, for the magnetic coupling problem, the flux $\mathcal{B}_{ij}^k$ is caused by the magnetic field that is not localized. Therefore, we still need non-local resistances to model the long-range effect of inductance accurately. Hence, the KCL equation (7.14) in our method is related not only to the localized $\hat{R}_{ij}^k$ $(j \in n_i)$, but also to all other $\hat{R}_{ij}^k$ $(j \neq i, j \in N)$. The experimental results in Subsection 7.3.1 will show that compared with the PEEC model, the full VPEC model considering all filaments is more accurate than the localized VPEC model from [86].

7.2 VPEC via PEEC inversion

Because of the difficulty of determining $\mathcal{B}_{ij}^k$ explicitly, there is no efficient calculation method for effective resistances in [86]. In this part, we will derive the circuit-level system equation based on the VPEC effective resistance matrix $\hat{G}$, and then present an efficient method to calculate effective resistances from the inversion of the partial inductance matrix.

We take the time derivative at both sides of (7.14) and then use (7.15) to replace the time derivative of the vector potential. Consequently, we obtain:

$$\left(\frac{1}{\hat{R}_{i0}^k} + \sum_{j\neq i} \frac{1}{\hat{R}_{ij}^k}\right)V_i^k + \sum_{j\neq i}\left(-\frac{1}{\hat{R}_{ij}^k}\right)V_j^k = l^2 \frac{\partial I_i^k}{\partial t}. \tag{7.18}$$

We define the circuit matrix $\hat{G}$ of VPEC model,

$$\hat{G}_{ij}^k = -\frac{1}{\hat{R}_{ij}^k}, \qquad \hat{G}_{ii}^k = \left(\frac{1}{\hat{R}_{i0}^k} + \sum_{j\neq i} \frac{1}{\hat{R}_{ij}^k}\right). \tag{7.19}$$

Then, the system equations can be rewritten as:

$$\hat{G}_{ii}^k V_i^k + \sum_{j \neq i} \hat{G}_{ij}^k V_j^k = l^2 \frac{\partial I_i^k}{\partial t}. \tag{7.20}$$

Compared with the system equation based on inductance matrix L^k and its inverse $S^k = (L^k)^{-1}$:

$$\begin{aligned} L_{ii}^k \frac{\partial I_i^k}{\partial t} + \sum_{j \neq i} L_{ij}^k \frac{\partial I_j^k}{\partial t} &= V_i^k, \\ S_{ii}^k V_i^k + \sum_{j \neq i} S_{ij}^k V_j^k &= \frac{\partial I_i^k}{\partial t}, \end{aligned} \tag{7.21}$$

we find that $\hat{G}^k$ and S^k only differ by a geometrical factor l^2:

$$\hat{G}^k = l^2 S^k. \tag{7.22}$$

Therefore, starting with the L matrix under the PEEC model, we first obtain the inverse of L, and then have the following extraction formula for $\hat{R}$ under the VPEC model:

$$\hat{R}_{ij}^k = -\frac{1}{l^2 S_{ij}^k}, \qquad \hat{R}_{i0}^k = \frac{1}{l^2 S_{ii}^k + \sum_{j \neq i} l^2 S_{ij}^k}. \tag{7.23}$$

Because the major computation effort is the inversion of the L matrix, we call this method an *inversion*-based VPEC model, which leverages the existing PEEC extractor. In contrast, the localized VPEC model is *integration* based and it needs to calculate explicitly the local flux $\mathcal{B}_{ij}^k$ $(j \in n_i)$ from scratch, where its accuracy is sensitive to the size of the control volume during the integration [86]. Therefore, it is highly accurate only for a small number of filaments and needs to optimize the size of the control volume for each filament when the system has a large number of filaments. Clearly, it becomes impractical for the full-chip extraction.

7.2.1 *Magnetic energy in the VPEC model*

Generally, the magnetic energy is given by the following space integral [57]:

$$\begin{aligned} u_m &= \frac{1}{2} \int_\tau \mathbf{A} \cdot \mathbf{J} d\tau \\ &= \sum_{k=x,y,z} u_m^k. \end{aligned} \tag{7.24}$$

For the full VPEC model, (7.24) can be rewritten by:

$$
\begin{aligned}
u_m^k &= \frac{1}{2}\int_{\mathcal{T}} A^k J^k \mathrm{d}\mathcal{T} \\
&= \frac{1}{2}\sum_i A_i^k \cdot l \cdot I_i^k \\
&= \frac{1}{2}\sum_i A_i^k \hat{I}_i^k.
\end{aligned}
\tag{7.25}
$$

Furthermore, when we rewrite the KCL equation (7.10) in terms of $\hat{G}$ matrix,

$$\sum_j G_{ij}^k A_i^k = \hat{I}_i^k, \tag{7.26}$$

we have the following relation for the magnetic energy under the full VPEC model:

$$u_m^k = \frac{1}{2}\sum_{i,j} G_{ij}^k A_i^k A_j^k. \tag{7.27}$$

Below we prove that the $\hat{G}$ matrix is positive definite.

7.2.2 *Property of $\hat{G}$ matrix*

Theorem 7.1. *Circuit matrix $\hat{G}^k$ $(k = x, y, z)$ in the VPEC model is positive definite.*

Because $\hat{G}$ only differs from the K matrix by a positive geometric constant, the proof of the matrix property (passivity and strict diagonal dominance) for K is equivalent for $\hat{G}$. The existing proofs in [16, 25] are based on the analogy: $[L] = \mu\epsilon[C^{-1}]$, which holds when $[C][L] =$ constant. However, this relation does not hold in general, as shown in [55]. Below, we present a direct proof for the VPEC model.

Proof: According to (7.27), because the energy u_m^k $(k = x, y, z)$ is positive, it automatically results in a positive definite matrix $\hat{G}^k$ [128].

Therefore, the corresponding VPEC model is passive. However, to further guarantee a passive model after truncating small-valued off-diagonal elements from the original positive definite matrix, we will prove that the matrix $\hat{G}$ is *strictly diagonal dominant* [128], i.e., $\hat{G}_{ii}^k > \sum_j |\hat{G}_{ij}^k|$.

Lemma 7.1. *All the effective resistances $\hat{R}_{i0}^k$ and $\hat{R}_{ij}^k$ $(k = x, y, z)$ in the VPEC model are positive.*

Proof: We present the proof based on the KCL equation (7.10). Since effective resistances are only determined by the geometry of the filaments, it will not depend on the applied external sources. Without loss of generality, we assume that an impulse current I_i^k is applied at filament a_i along the z direction, and all other filaments a_j are connected to the vector potential ground. Note that for filament a_i, its average vector potential A_i^k is in the same direction as I_i^k; for any other grounded filament a_j, its average vector potential A_j^k is zero, but its *induced* current $-I_j^k$ is in the *opposite* direction to I_i^k according to *Lenz's law.* Hence for filament a_j, (7.10) becomes

$$\frac{(A_j^k - A_i^k)}{\hat{R}_{ij}^k} = \frac{-A_i^k}{\hat{R}_{ij}^k} = -lI_j^k,$$

where the induced current I_j^k is determined by the coupling flux between a_i and a_j. Equation (7.28) can be further rewritten by:

$$\hat{R}_{ij}^k = \frac{A_i^k}{lI_j^k} > 0. \tag{7.28}$$

The positiveness of the ground resistance $\hat{R}_{i0}^k$ can be easily proved in a similar fashion. With this Lemma, we can further prove the following Theorem:

Theorem 7.2. *Circuit matrix $\hat{G}^k$ ($k = x, y, z$) in the VPEC model is strictly diagonally dominant.*

Proof: According to (7.19) we have

$$\sum_{j \neq i} |\hat{G}_{ij}^k| = \sum_{j \neq i} \frac{1}{\hat{R}_{ij}^k} \tag{7.29}$$

and

$$\sum_{j \neq i} |\hat{G}_{ij}^k| < \frac{1}{\hat{R}_{i0}^k} + \sum_{j \neq i}^{N} \frac{1}{\hat{R}_{ij}^k} = \hat{G}_{ii}^k,$$

or

$$\hat{G}_{ii}^k > \sum_{j \neq i} |\hat{G}_{ij}^k|. \tag{7.30}$$

I.e., the circuit matrix $\hat{G}$ is strictly diagonally dominant. Note that truncating small off-diagonal entries from a strictly diagonally dominant matrix still leads to a positive definite matrix, i.e., a passive circuit model [128]. Based on Theorem 7.2, such a truncation-based sparsification still leads to passive circuit models. Intuitively, truncating small off-diagonal entries in the $\hat{G}$ matrix (equivalent to truncating larger off-diagonal entries in $\hat{R}$ matrix) results in ignoring larger resistors in the equivalent resistance network. Because the larger resistors are less sensitive to and also contribute less to the current change, the resulting sparsified model can still have a good waveform accuracy, as presented in Subsection 7.3.2. Moreover, our proof assumes that wires can be decomposed into short wires with a similar length. Therefore, in our experiments, we always segment wires according to one-tenth of maximum operating frequency when the wire lengths are different.

7.3 Numerical examples

7.3.1 *Simulation of aligned parallel bus*

We consider a five-bit bus, with one-segment per line. Each bus line is 1000 μm long, 1 μm wide and 1 μm thick. The space between lines is 2 μm. With the extracted RLCM parameters under the PEEC model, we further construct the full VPEC model (with coupling $\hat{R}_{ij}$ between all bits) from this method, and the localized VPEC model (with coupling $\hat{R}_{ij}$ between adjacent bits) from [86]. We measured responses at the far end of all five bits, and compared waveforms of the second bit in Figure 7.3 (a)–(b), for both the time-domain and the frequency-domain simulations, where a 1-V step voltage with 10 ps rising time is used for the time-domain transient simulation, and a 1-V AC voltage for frequency-domain (1 Hz $\sim$ 10 GHz) simulation. Clearly, the full VPEC model and the PEEC model produce identical waveforms in both frequency-domain and time-domain simulations, but the localized VPEC model introduces a non-negligible error and is not accurate compared with the PEEC model, where the time-domain response shows 15% waveform difference and the frequency-domain response shows a large deviation beyond 5 GHz.

7.3.2 *Truncated-VPEC model*

We further present the truncated-VPEC (tVPEC) model. After the full inversion of L, we obtain a strictly diagonally dominant matrix $\hat{G}$. As explained in Section 7.2, its small-valued off-diagonal elements can be truncated without loss of passivity. We present two truncating approaches below: geometrical (gtVPEC) and numerical (ntVPEC) truncation. The first one is applicable to the aligned parallel bus, and the second is applicable to conductors of any shapes.

Geometrical truncation

For the aligned parallel bus, we can define a *truncating window* (N_W, N_L) for each wire segment, where N_W and N_L are the numbers of coupled segments in directions of wire width and length, respectively. The coupling along the wire length is the *forward* coupling, and the one along the wire width is the *aligned* coupling. Because of the symmetry introduced by aligning and paralleling, each wire segment will have the same-sized truncating window. As a result, the tVPEC model only contains $\hat{R}_{ij}$

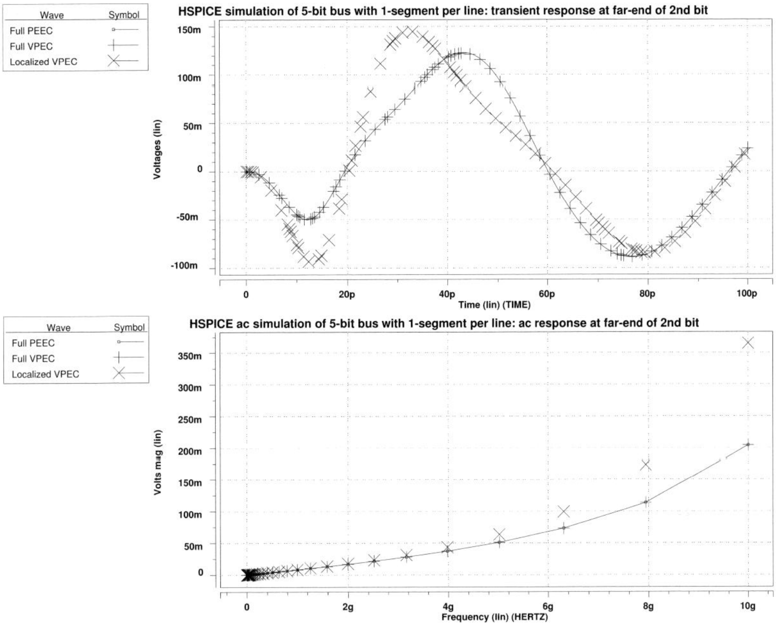

Figure 7.3 For five-bit bus, (a) a 1-V step voltage with 10 ps rising time and (b) a 1-V *ac* voltage are applied to the first bit and all other bits are quiet. The responses of the PEEC model, full VPEC model, and localized VPEC model are measured at the far end of the second bit. Reprinted with permission from [135] (c) 2005 IEEE.

within the truncating window for each wire segment, and is called *gt*VPEC in Table 7.2.

We consider a 32-bit bus with eight-segments per line and four differently sized truncating windows: (32, 8), (32, 2), (16, 2), and (8, 2), and summarize the experiment settings and results in Table 7.2. Clearly, there is a smooth trade-off between runtime and accuracy for different truncating window sizes, where the average voltage differences and associated with standard deviations are calculated for the entire time steps in SPICE simulation. We first compare results of different truncating windows. The truncating window (8, 2) achieves the highest speed-up of 30× and the largest difference of about 0.2 mV on average, less than 2% of the noise peak, and the truncating window (32, 2) has the highest accuracy with 0.06 mV on average but a reduced speed-up of 10×. Furthermore, the small difference between windows (32, 8) and (32, 2) implies that the forward couplings between non-adjacent segments are negligible. However, an N_W larger than N_L (as shown in Table 7.2) is needed to achieve a high accuracy. This implies that the aligned coupling is stronger than the forward coupling and, considering only the adjacent aligned coupling may lead to a large error.

Models and window sizings	Number of elements	Run time (s) and speed-up	Average voltage difference (V)	Standard deviation (V)
Full PEEC	32896	2535.48 (1×)	0	0
Full VPEC(32, 8)	32896	772.89 (3×)	1.00×10^{-5}	6.26e-4
*gt*VPEC(32, 2)	11392	311.22 (8×)	5.97×10^{-5}	1.84×10^{-3}
*gt*VPEC(16, 2)	3488	152.57 (16×)	-1.23×10^{-4}	4.56×10^{-3}
*gt*VPEC(8, 2)	2240	85.14 (32×)	-2.17×10^{-4}	8.91×10^{-3}

Table 7.2 *Settings and results of geometrical tVPEC models. Reprinted with permission from [135] (c) 2005 IEEE.*

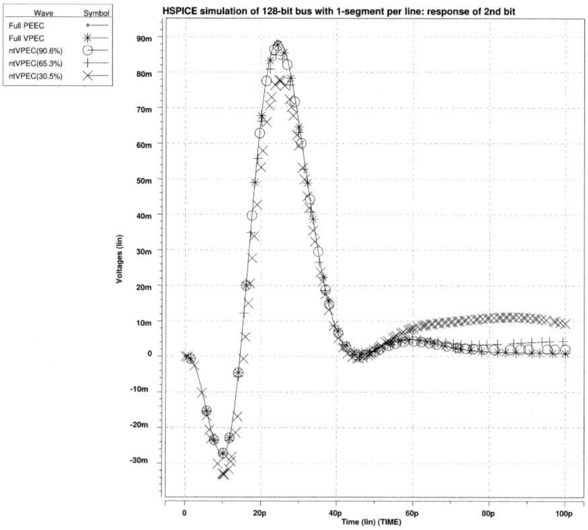

Figure 7.4 For 128-bit bus by numerical truncation, a 1-V step voltage with 10 ps rising time is applied to the first bit, and all other bits are quiet. The responses of the PEEC model, the full VPEC model, and the *t*VPEC model are measured at the far end of the second bit. Reprinted with permission from [135] (c) 2005 IEEE.

Numerical truncation

For the numerical truncation, we define the *coupling strength* as the ratio of an off-diagonal element to its correspondent diagonal element at each row of $\hat{G}$. We then truncate those off-diagonal elements with coupling strength smaller than a specified threshold.

Figure 7.4 plots the simulation results under the numerical sparsification for the non-aligned parallel bus with 128 bits and one segment per line. The sparse factor is the ratio between the numbers of circuit elements in the truncated and full VPEC models. The waveform difference is small in terms of the noise peak for sparse factors up to 30.5%. Table 7.3 summarizes the truncation setting and simulation result, where values in parentheses of column 1 are truncating thresholds, and the runtime includes both SPICE simulation and matrix inversion in the case of VPEC models. One can see from the table that up to 30× speed-up is achieved when the average waveform differences is up to 0.377 mV, less than 1% of the noise peak.

Models and truncation thresholds	Number of elements	Runt time (s) and speed-up	Average voltage difference (V)	Standard deviation (V)
Full PEEC	8256	281.02 (1×)	0	0
Full VPEC(0.0)	8256	36.40 (7×)	-1.64×10^{-6}	3.41×10^{-4}
ntVPEC(5×10^{-5})	7482	30.89 (9×)	4.64×10^{-6}	4.97×10^{-4}
ntVPEC(1×10^{-4})	5392	19.55 (14×)	1.29×10^{-5}	1.37×10^{-3}
ntVPEC(5×10^{-4})	2517	8.35 (28×)	3.77×10^{-4}	5.20×10^{-3}

Table 7.3 *Settings and results of numerical tVPEC models. Reprinted with permission from [135] (c) 2005 IEEE.*

A larger speed-up factor can be expected, as a higher waveform difference can be tolerated in practice. We also compare the full VPEC model and the PEEC model. The full VPEC simulation is 7× faster and has a negligible waveform difference.

7.4 Inductance models in hierarchical reduction

In this section, we discuss the VPEC (vector potential equivalent circuit) used in hierarchical reduction. As a continued discussion, we further compare the inductance formulation by the VPEC model and by the nodal susceptance methods [9,17,98]. We will show that the inductance model by nodal susceptance is not physically equivalent to inductance as unwanted DC paths are created at low frequency.

7.4.1 *Inductance formulation by nodal analysis*

For an RLCM circuit, when we assume that only independent current sources exist at external ports, the circuit matrix in the frequency domain starting with MNA formulation can be written as

$$\mathcal{G}x + s\mathcal{C}x = Bi(s), \qquad v(s) = B^T x, \tag{7.31}$$

where x, v, and i are the state variables, output voltage and input current vectors, and $\mathcal{G}, \mathcal{C}$, and B are the state and the input and output matrices, respectively. Equation (7.31) can be further written as:

$$\begin{bmatrix} G & A_l^T \\ -A_l & 0 \end{bmatrix} \begin{bmatrix} v_n \\ i_l \end{bmatrix} + s \begin{bmatrix} C & 0 \\ 0 & L \end{bmatrix} \begin{bmatrix} v_n \\ i_l \end{bmatrix} = \begin{bmatrix} A_i i_n(s) \\ 0 \end{bmatrix}, \tag{7.32}$$

where G and C are the admittance matrices for resistors and capacitors, L is the inductance matrix, which includes the mutual inductance; v_n is a vector of node voltage; i_l is a branch current vector of inductors; A_l is the adjacency matrix for all inductors; and A_i is the adjacency matrix for all port current sources.

Circuit reduction means applying the Gaussian elimination for state variables like node voltage v_n and branch current i_l. If we first reduce the branch current vector i_l, we actually result in the state equation only with the nodal voltage variables

$$[G + sC + \frac{1}{s}\Gamma][v_n] = [A_i i_n(s)], \tag{7.33}$$

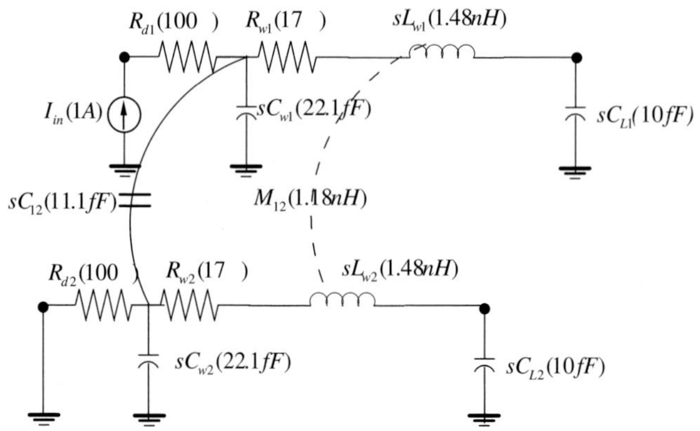

Figure 7.5 Example of a coupled two-bit RLCM circuit under the PEEC model. Reprinted with permission from [135] (c) 2005 IEEE.

where $\Gamma = A_l S A_l^T = A_l L^{-1} A_l^T$ is the *nodal susceptance* [9, 17, 98]. This is exactly the nodal analysis formulation with Γ representing the inductance. The circuit-reduction by further eliminating the nodal voltage-variable v_n is exactly the Y-Δ transformation in [98].

The nodal susceptance in (7.33) actually creates DC paths at low-frequency range. We illustrate this with a two-bit interconnect example shown in Figure 7.5. The nodal voltage equation of susceptance at four nodes (A, B, C, D) becomes

$$\begin{aligned}
(S_{11}/s)V_A - (S_{11}/s)V_B + (S_{21}/s)V_C - (S_{21}/s)V_D &= I_1 \\
-(S_{11}/s)V_A + (S_{11}/s)V_B - (S_{12}/s)V_C + (S_{12}/s)V_D &= -I_1 \\
(S_{12}/s)V_A - (S_{12}/s)V_B + (S_{22}/s)V_C - (S_{22}/s)V_D &= I_2 \\
-(S_{12}/s)V_A + (S_{12}/s)V_B - (S_{22}/s)V_C + (S_{22}/s)V_D &= -I_2.
\end{aligned} \tag{7.34}$$

As shown in Figure 7.6, this is mathematically equivalent to stamping the six susceptance elements into the admittance matrix [98] when $s \neq 0$. However, when the susceptance element S_{ij}/s approaches infinite (thus 0 impedance or short circuit) when $s = 0$, there exist four unwanted DC-paths between nodes (A, B, C, D), which did not exist before. As a result, it leads to the wrong *dc* values and inaccurate low-frequency simulation results, even for the two-bit bus example shown in Figure 7.5.

We compute the exact driving-point impedance responses using inductance under MNA and the nodal susceptance under NA, respectively, using the symbolic analysis tool [110]. As shown in Figure 7.7, NA formulation (by using the nodal susceptance for inductance) gives the exact response as SPICE does in the high-frequency range, but the response is not correct in the low-frequency range. When s approaches zero, the actual driving-point impedance in Figure 7.5 should be dominated by three capacitors with a total capacitance value 43 fF. However, for Figure 7.6 at

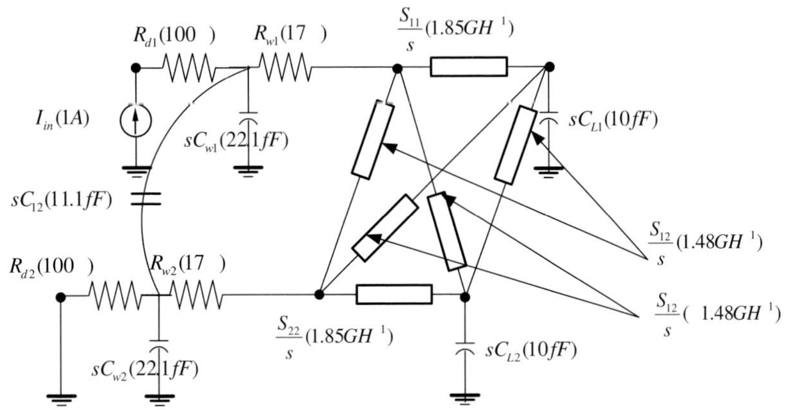

Figure 7.6 Example of a coupled two-bit RLCM circuit under the nodal susceptance model. Reprinted with permission from [135] (c) 2005 IEEE.

DC, the driving-point impedance becomes a resistor with a total resistance value 234 Ω(ohms) (or 49 Ω(dB)) owing to the unwanted DC paths.

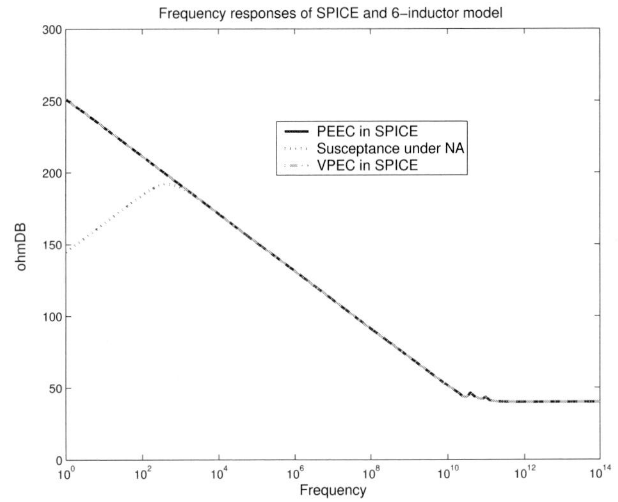

Figure 7.7 Frequency responses of PEEC model in SPICE, susceptance under NA and VPEC models for the two-bit bus. Reprinted with permission from [135] (c) 2005 IEEE.

There are two reasons for such a discrepancy:

- The nodal admittance Γ/s is indefinite at DC.
- More importantly, the Γ matrix itself is singular. This is due to the fact that the number of nodal-voltage variables n_v is usually much larger than the number of inductive branch-current variables n_i. There are originally only n_i independent variables for L, but the dimension of Γ becomes n_v, which is much larger than

	v1	v2	v3	v4	v5	v6
v1	g	-g				
v2	-g	g				
v3						
v4				g	-g	
v5				-g	g	
v6						

(a)

	v1	v2	v3	v4	v5	v6
v1						
v2		s	$-s$		s_x	$-s_x$
v3		$-s$	s		$-s_x$	s_x
v4						
v5		s_x	$-s_x$		s	$-s$
v6		$-s_x$	s_x		$-s$	s

(b)

	v1	v2	v3	v4	v5	v6
v1	c_x			$-c_x$		
v2						
v3			c			
v4	$-c_x$			c_x		
v5						
v6						c

(c)

	p1	p2
v1	1	
v2		
v3		
v4		1
v5		
v6		

(d)

Figure 7.8 Stamp of the second-order admittance in the NA matrix, where (a), (b) and (c) represent for G, Γ, C and B. G (rank=4) and Γ (rank=4) are both singular for 6×6 matrices. Reprinted with permission from [135] (c) 2005 IEEE.

n_i. Therefore, Γ is singular and has a low rank.

Figure 7.8 shows the stamping of Γ for the example of two wires in Figure 7.6. As a result, the NA formulation of inductance, which is based on L^{-1}, is no longer equivalent to the original circuit matrix. Hence circuit reduction starting with nodal-susceptance formulation cannot give the correct low-frequency response in general, and is not suitable for generating wideband macro-models of the interconnects.

7.4.2 *Inductance formulation by VPEC model*

From the above discussion, we know that inductance formulation by nodal susceptance leads to an inaccurate low-frequency response. It is not suitable for generating reduced interconnect models for wideband applications. However, directly handling mutual inductance in a dense MNA formulation, as in [3], will be computationally expensive. As shown in [134], the sparsified VPEC model actually not only achieves the runtime speedup, but also has high accuracy compared with the original full model. Therefore, we use the VPEC model to represent inductance in our circuit-reduction flow, as it enables passive pre-sparsification [86, 134].

The significant difference between VPEC and nodal-susceptance models for mutual inductance is that VPEC is a physically equivalent model, and it can exactly represent the original system [134]. As shown in Figure 7.9, this model consists of an

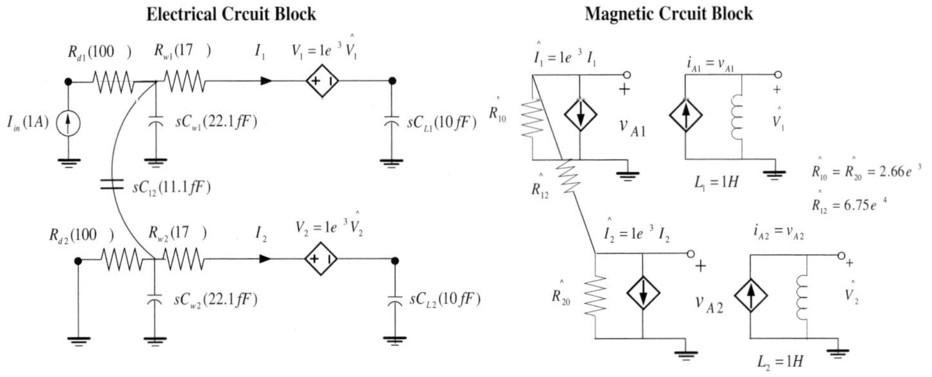

Figure 7.9 Example of a coupled two-bit RLCM circuit under the VPEC model. Reprinted with permission from [135] (c) 2005 IEEE.

electrical circuit (PEEC resistance and capacitance) and magnetic circuit (VPEC effective resistance and controlled source). It includes the following components: (1) the wire resistance and the capacitance are the same as in the PEEC model; (2) a dummy voltage source (sensing electrical current I_i) to control $\hat{I}_i$; (3) a voltage-controlled current source to relate $\hat{V}_i$ and $\hat{I}_i$ with gain $g = 1$; (4) an electrical voltage source V_i controlled by $\hat{V}_i$; (5) effective resistors including ground $\hat{R}_{i0}$ and coupling $\hat{R}_{ij}$ to consider the strength of inductances; and (6) a unit inductance L_i to: (i) take into account the time derivative of A_i; and (ii) preserve the magnetic energy from the electronic circuit.

Clearly, this SPICE compatible implementation does not introduce unwanted DC paths when $s = 0$ as by the nodal-susceptance. Moreover, Figure 7.7 shows the response of the VPEC model for the 2-bit circuit, which is identical to SPICE for the entire frequency range. Detailed analysis also shows that the impedance function of the 2-bit circuit modeled by the VPEC model is the same as the impedance function by PEEC model.

As shown in [134], although the VPEC model introduces more circuit elements, it has a faster runtime because this model dramatically reduces the number of reactive elements (i.e., inductors), and leads to fewer numerical derivatives and integrals and makes the simulation converge faster. To further improve the sparsified VPEC model extraction without full inversion, as in [134], we extend a windowing technique [9]. It reduces the computation complexity to $(O(Nb^3))$, where b is the size of the window (i.e., the size of the submatrix). Note that although the VPEC model enables efficient inductance simulation, the order of the circuit matrix is still high. Moreover, its SPICE compatible model still contains controlled sources, and cannot be handled by the existing realizable circuit-reduction approaches [3, 98].

7.5 Summary

The primary content of this chapter is to derive the inversion-based full VPEC model for multiple inductive interconnects and illustrate how to build sparsified VPEC for SPICE simulation with guaranteed passivity.

Using a vector potential, we developed the VPEC model to model the inverse-inductance elements. The resulting VPEC circuit is passive and has a strictly diagonally dominant matrix. This enables truncation-based sparsification methods with guaranteed passivity. We have presented the truncation-based method and have achieved orders of magnitude speedup in circuit simulation with small errors compared to the PEEC model. We have also shown that the matrix $\hat{G}$ can be used to justify the K-element or susceptance based simulation [9,16,25,58,141] from first principles. Note that SPICE can directly simulate the VPEC model but not the K-element based model. In addition, we also showed that the direct circuit stamping of the K elements by NA leads to singular matrices. In contrast, the VPEC model can be accurately stamped in SPICE.

8 Structure-preserving model order reduction

8.1 Introduction

The integrated circuit industry has continuously enjoyed enormous success owing to its ever increasing large-scale integration. With the advent of system-on-a-chip (SOC) technology [30,133], it requires heterogeneous integration to support different modules within one single silicon chip such as logic, memory, analog, RF/microwave, FPGA, and MEMS sensor. Such a heterogeneous integration leads to highly non-uniform current distribution across one layer or between any pair of layers. As a result, it is beneficial to design a structured multi-layer and multi-scale power and ground (P/G) grid [11] that is globally irregular and locally regular [115] according to the current density. This results in a heterogeneously structured P/G circuit model in which each subblock can have a different time constant. In addition, the typical extracted P/G grid circuits usually have millions of nodes and large numbers of ports. To ensure power integrity, specialized simulators for structured P/G grids are required to efficiently and accurately analyze the voltage bounce or drop using macro-models.

In [139], internal sources are first eliminated to obtain a macro-model with only external ports. The entire P/G gird is partitioned at and connected by those external ports. Because elimination results in a dense macro-model, [139] applies an additional sparsification procedure that is error-prone and inefficient. In addition, [18,95] proposed localized simulation and design methods based on the locality of the current distribution in most P/G grids with C4-bumps. The P/G grid is divided into several subblocks, where the current sources in each block only affect the voltage fluctuation in that block. Nevertheless, all these methods [18,95,139] do not consider the heterogeneous structure of P/G grids resulting from the non-uniform current density. It results in a flat time-domain transient simulation that all blocks use the same time-step. It would improve the efficiency if each partitioned block has a different time constant and, hence, can be simulated at a different time-step, as in the waveform relaxation method [132].

An alternative approach for obtaining a macro-model is to use moment matching [91] by projection-based model order reduction (MOR) [32, 63, 85, 113], where PRIMA [85] is the method widely used to efficiently generate an order-reduced macro-model with preserved passivity. The reduced model of PRIMA [85] by a projection matrix with order q can match $n = \lfloor q/n_p \rfloor$ block moments (where n_p is the port number). PRIMA can be implemented in a fashion of iterative path-tracing to efficiently solve *tree* structured P/G grids [120]. However, it is too inefficient to be directly applied to *mesh* structured P/G grids. The difficulty in appling MOR to analyze P/G grids stems mainly from the following reasons. The cost of Arnoldi orthonormalization is high for large circuits, and moment matching using the block Krylov subspace is less accurate with an increased number of ports. In addition, the reduced macro-model is dense, which slows down simulation when the port number is large [34]. To reduce orthonormalization cost for large circuits, HiPRIME [71] applies a partitioned PRIMA to reduce the entire circuit in a divide-and-conquer fashion. After gluing the reduced state matrices, HiPRIME performs an additional projection to further reduced the entire system. However, all these approaches [71,85] use a flat projection that leads to a loss in the structure information of the state matrices. For example, the original state matrices may be sparse, but they become dense after flat projection. The resulting macro-model, therefore is too dense to be efficiently factorized in the time-domain or frequency-domain simulation. To leverage the subblock structure in $\mathbf{G}$ and $\mathbf{C}$ matrices during macro-model construction, we introduce a *structured model order reduction* method in this chapter.

8.2 Chapter overview

In Section 8.3, we first review the moment-matching theorems. Instead of matching block moments of the transfer function, we directly match moments of output with an *excitation current vector*. As a result, the first q moments or q dominant poles of output can be matched using a projection matrix with order q, which is independent of port number. In contrast, the number of matched block moments by PRIMA decreases as the port number increases.

In Section 8.4, we first review a recent 2×2 structure-preserving model order reduction: SPRIM [37]. We further introduce a generalized block structure-preserving model order reduction (BSMOR). Unlike SPRIM, it further partitions the flat projection matrix obtained from PRIMA into m blocks according to the provided block representation of state matrices. We show that this reduction preserves the block structure. As a result, the macro-model obtained by BSMOR is intrinsically sparse, and does not need the LP-sparsification used in [139]. Moreover, we show that the reduced macro-model can exactly match q dominant and additional $(m-1)q$ approximated poles. However, these two methods do not consider the heterogeneous substructure of state matrices. The reduced model is not compact or optimum, as discussed in Section 8.5 later on. In addition, as in [85], it orthonormalizes the entire state matrices to obtain the projection matrix. As a result, it is inefficient for large sized circuits.

In Section 8.5, we propose a triangularization-based structure-preserving model order reduction (TBS) method.

- We first represent the original system by interconnected *basic blocks*. The basic blocks are obtained from the current density of locally regular structures in P/G grids. We reduce each basic block *independently* with order q, determine its first q dominant poles, and obtain its corresponding projection matrix.
- We then carry out dominant-pole-based clustering to obtain m clusters of basic blocks, where m is decided by the nature of the structured network. Each cluster is called a *compact block* with a unique pole distribution and a projection matrix. Because clustering reduces the redundant block information, the block-based form in our method is more compact than that in BSMOR.
- We further triangulate the system into a triangular system with m compact blocks in the diagonal. The poles of the resulting triangular system are determined only by m diagonal blocks. A block-diagonal-structured projection matrix is constructed by stacking projection matrices for individual diagonal blocks in the triangular system. The reduced triangular system can be proved to match mq poles of the original one. This is the *primary contribution* of this method. Because PRIMA or HiPRIME can only match q poles using the same number of moments, the reduced system by TBS is more accurate, or TBS has a higher reduction efficiency under the same error bound. Moreover, the mq poles are exactly matched in TBS but not in BSMOR.

In addition, as discussed in Section 8.6, the obtained macro-model by BSMOR and TBS also enables a two-level analysis similar to [47] to solve each reduced block independently with different time steps. It reduces the transient simulation time in both frequency and time domains. In contrast, the reduced model by PRIMA and HiPRIME is dense and cannot be analyzed in a fashion of two-level analysis. We present the experiments in Section 8.7.

8.3 Background

8.3.1 *Grimme's moment-matching theorem*

Using modified nodal analysis (MNA), the system equation of a P/G grid in the frequency domain is

$$(\mathbf{G} + s\mathbf{C})\mathbf{x}(s) = \mathbf{B}u(s), \quad \mathbf{y}(s) = \mathbf{B}^T\mathbf{x}(s) \tag{8.1}$$

where $x(s)$ is the state variable vector, $\mathbf{G}$ and $\mathbf{C}$ ($\in R^{N\times N}$) are state matrices for conductance and capacitance with size N, and $\mathbf{B}$ and $\mathbf{L}$ ($\in R^{N\times n_p}$) are input and output port-incident matrices with n_p ports.

Eliminating $x(s)$ in (8.1) gives

$$H(s) = \mathbf{L}^T(\mathbf{G} + s\mathbf{C})^{-1}\mathbf{B}, \tag{8.2}$$

$H(s)$ is a multiple-input multiple-output (MIMO) transfer function. PRIMA [85] finds a projection matrix V ($\in R^{N\times n}$). It has dimension q and its columns span

n-block ($n = \lceil q/n_p \rceil$) Krylov subspace $\mathcal{K}(\mathbf{A}, \mathbf{R}, n)$, i.e,

$$\mathcal{K}(\mathbf{A}, \mathbf{R}, n) = \text{span}(V) = \{\mathbf{R}, \mathbf{AR}, ..., \mathbf{A}^{n-1}\mathbf{R}\}, \tag{8.3}$$

where two *moment-generating matrices* are

$$\mathbf{A} = (\mathbf{G} + s_0\mathbf{C})^{-1}\mathbf{C}, \quad \mathbf{R} = (\mathbf{G} + s_0\mathbf{C})^{-1}\mathbf{B},$$

and s_0 is the expansion point that ensures that $\mathbf{G} + s_0\mathbf{C}$ is non-singular. The reduced transfer function is

$$\hat{H}(s) = \hat{\mathbf{L}}^T(\hat{\mathbf{G}} + s\hat{\mathbf{C}})^{-1}\hat{\mathbf{B}}, \tag{8.4}$$

where

$$\hat{\mathbf{G}} = V^T\mathbf{G}V, \ \hat{\mathbf{C}} = V^T\mathbf{C}V, \ \hat{\mathbf{B}} = V^T\mathbf{B}, \ \hat{\mathbf{L}} = V^T\mathbf{L}.$$

Note that $\hat{\mathbf{G}}$ and $\hat{\mathbf{C}} \in R^{q \times q}$, and $\hat{\mathbf{B}}$ and $\hat{\mathbf{L}} \in R^{q \times n_p}$. As proved in [45], $\hat{H}(s)$ preserves the block moments of $H(s)$, i.e.,

Theorem 8.1. *If $\mathcal{K}(\mathbf{A}, \mathbf{R}, n) \subseteq span(V)$, then the first n expanded block moments at s_0 are identical for $\hat{H}(s)$ and $H(s)$.*

8.3.2 Moment matching of output response

According to Theorem 8.1, if there is only one port, i.e., we have a (single-input single-output) SISO system, the reduced model can match q moments. When the port number n_p is large, which is typical for P/G grids, the order of matched block moment n is reduced and the reduced transfer function $\hat{H}(s)$ is less accurate. In this case, it is better to define an *excitation current vector* $\mathbf{J} = \mathbf{B}u(s)$ to directly match the moment of output $\mathbf{x}(s) = (\mathbf{G} + s\mathbf{C})^{-1}\mathbf{J}$ with the input vector $\mathbf{J}$. As a result, the matched moments of the output with input $\mathbf{J}$ are q, which is independent of the port number n_p. This is because a MIMO system with right-hand-side $\mathbf{B}u$ can be transformed into superposed SISO systems with the input $\mathbf{J}$. The following Theorem has been proved.

Theorem 8.2. *Assume a MIMO system with a unit-impulse current source u, and define the excitation current vector $\mathbf{J} = \mathbf{B}u$, where $u \in R^p$ and $\mathbf{J} \in R^N$. When the q columns of projection matrix V are obtained, the reduced response at the output $\hat{\mathbf{x}}(s) = (\hat{\mathbf{G}} + s\hat{\mathbf{C}})^{-1}\hat{\mathbf{J}}$ ($\hat{\mathbf{J}} = V^T\mathbf{J}$) matches the first q moments of the original $\mathbf{x}(s) = (\mathbf{G} + s\mathbf{C})^{-1}\mathbf{J}$.*

Note that the following two systems have the same output $x(s)$

$$(\mathbf{G} + s\mathbf{C})\mathbf{x}(s) = \mathbf{B}u(s), \quad (\mathbf{G} + s\mathbf{C})\mathbf{x}(s) = \mathbf{J}(s). \tag{8.5}$$

In addition, $\mathbf{J}$ can be decomposed into several excitation components

$$\mathbf{J} = \sum_{i=1}^{p} \mathbf{J}_i = [J_1 \quad 0 \quad ... \quad 0]^T + ... + [0 \quad ... \quad J_p \quad 0]^T,$$

Clearly for each $\mathbf{J}_i$, it is equivalent to excite a SISO system with input $\mathbf{J}_i$. Therefore, $\hat{\mathbf{x}}_i(s)$ matches the first q moments of $\mathbf{x}_i(s)$. With superposition, it is easy to verify that $\sum_{i=1}^{p} \hat{\mathbf{x}}_i(s)$ matches the first q moments of $\sum_{i=1}^{p} \mathbf{x}_i(s)$. In contrast, PRIMA [85] matches the block moment of transfer function with input matrix $\mathbf{B}$.

Moreover, we have

Corollary 8.1. *With the input* $\mathbf{J}$*, the first* q *dominant poles of* $\mathbf{x}(s)$ *are matched by* $\hat{\mathbf{x}}(s)$.

Using excitation current vector J as input, the first q moments are identical for $\mathbf{x}(s)$ and $\hat{\mathbf{x}}(s)$. So are the first q dominant poles.

Because the typical P/G grids contain large numbers of ports, in this chapter the MOR is performed to match the moment of output $\mathbf{x(s)}$ with the input $\mathbf{J} = \mathbf{B}u$, similar to [71,130]. This theorem can easily be extended to inputs with non-impulse current sources by using a *generalized excitation current source* with an augmented Arnoldi orthonormalization [112].

8.4 Block-structure-preserving model reduction

8.4.1 SPRIM method

The SPRIM method [37] observes a 2×2 block structure of the MNA matrix

$$\mathbf{G} = \begin{bmatrix} G & A^T \\ -A & 0 \end{bmatrix}, \mathbf{C} = \begin{bmatrix} C & 0 \\ 0 & L \end{bmatrix}, \tag{8.6}$$

where G ($\in R^{n_1 \times n_1}$), C ($\in R^{n_1 \times n_1}$), L ($\in R^{n_2 \times n_2}$) are the conductance, capacitance and inductance matrices, respectively, and A ($\in R^{n_2 \times n_1}$) is the adjacent matrix indicating the branch current flow at the inductor. Note that $n_1 + n_2 = N$.

It constructs a structured projection matrix $\widetilde{V}$ accordingly by partitioning the flat V obtained from the q-th PRIMA iteration

$$V = \begin{bmatrix} V_1 \\ V_2 \end{bmatrix} \rightarrow \widetilde{V} = \begin{bmatrix} V_1 & 0 \\ 0 & V_2 \end{bmatrix}. \tag{8.7}$$

where V_1 is $\in R^{n_1 \times qn_p}$, V_2 is $\in R^{n_2 \times qn_p}$, and $\widetilde{V}$ is $\in R^{N \times 2qn_p}$. As a result, the number of columns in $\widetilde{V}$ is twice that in V. Accordingly, the new reduced state matrices are

$$\widetilde{\mathbf{G}} = \begin{bmatrix} \widetilde{G} & \widetilde{A}^T \\ -\widetilde{A} & 0 \end{bmatrix}, \widetilde{\mathbf{C}} = \begin{bmatrix} \widetilde{C} & 0 \\ 0 & \widetilde{L}, \end{bmatrix}, \tag{8.8}$$

where

$$\widetilde{G} = V_1{}^T G V_1 \quad \widetilde{A} = V_2{}^T A V_1 \quad \widetilde{C} = V_1{}^T C V_1 \quad \widetilde{L} = V_2{}^T L V_2.$$

Similarly, the size of $\widetilde{\mathcal{G}}$, $\widetilde{\mathcal{C}}$ ($\in R^{2qn_p \times 2qn_p}$), and $\widetilde{\mathcal{B}}$ ($\in R^{2qn_p \times n_p}$) is twice than that of $\hat{\mathbf{G}}$, $\hat{\mathbf{C}}$, and $\hat{\mathbf{B}}$ reduced by using V. Therefore, the reduced model with state matrices has $\widetilde{\mathcal{G}}$ and $\widetilde{\mathcal{C}}$ approximately twice as much poles as those of the reduced model with state matrices $\hat{\mathcal{G}}$ and $\hat{\mathcal{C}}$. An improved accuracy over PRIMA is observed [37].

Since the reduced model is written in the first order form in (8.8), the model reduced by SPRIM is twice as large as that produced by PRIMA. But the reduced model produced by SPRIM preserves the structure of the original model and can be further reduced into the second-order form using Schur's decomposition [44] to element branch-current variable:

$$\widetilde{\mathcal{Y}}_{NA} = \widetilde{G} + s\widetilde{C} + \frac{1}{s}\widetilde{A}^T \widetilde{L}^{-1} \widetilde{A},$$

where $\widetilde{\mathcal{Y}}_{NA}$ is the reduced state matrix in NA form, which has the same size of the reduced matrix by using V. But the difference is that each element in $\widetilde{\mathcal{G}}_{NA}$ becomes a second-order rational function of s instead of first-order of s.

Hence, SPRIM algorithm essentially consists of two reduction steps: the first step is the structure-preserving projection-based reduction and the second step is block elimination. As a result, SPRIM can match more poles than PRIMA, and they result in the same *size* of the reduced model.

8.4.2 Block structured projection

Nevertheless, SPRIM does not consider the possible subblock structure inside the $\mathbf{G}$ and $\mathbf{C}$ matrices. For the structured system, the original P/G grids can be partitioned into m_0 *basic blocks*, where a dense grid with a small pitch is used for a region with high current density, and a sparse grid with a large pitch is used for a region with low current density [115]. Because of the heterogeneous structure of grids, each basic block can have different RC values, $\mathbf{G}_{ii}$ and $\mathbf{C}_{ii}$, which are interconnected by the *coupling blocks* $\mathbf{G}_{ij}$ and $\mathbf{C}_{ij}$ ($i \neq j$), respectively. As a result, the conductance matrix $\mathbf{G}$ can be written in a block representation

$$\mathbf{G} = \begin{bmatrix} \mathbf{G}_{1,1} & \mathbf{G}_{1,2} & \cdots & \mathbf{G}_{1,m_0} \\ \mathbf{G}_{2,1} & \mathbf{G}_{2,2} & \cdots & \mathbf{G}_{2,m_0} \\ \vdots & \vdots & \ddots & \vdots \\ \mathbf{G}_{m_0,1} & \mathbf{G}_{m_0,2} & \cdots & \mathbf{G}_{m_0,m_0} \end{bmatrix}, \tag{8.9}$$

where each block has the size n_k ($\sum_{k=1}^{m_0} n_k = N$). A similar block structure is assumed for the $\mathbf{C}$ matrix and $\mathbf{J}$ becomes

$$\mathbf{J} = \begin{bmatrix} \mathbf{J}_1 \ \mathbf{J}_2 \ \ldots \ \mathbf{J}_{m_0} \end{bmatrix}^T \quad \mathbf{x} = \begin{bmatrix} \mathbf{x}_1 \ \mathbf{x}_2 \ \ldots \ \mathbf{x}_{m_0} \end{bmatrix}^T, \tag{8.10}$$

where each block contains user-specified n_{p_k} ports ($n_p = \sum_{k=1}^{m_0} n_{p_k}$).

Accordingly, we further partition the qth order projection matrix V obtained from PRIMA according to the m_0 blocks in (8.9)

$$V = \begin{bmatrix} V_1 \\ V_2 \\ \vdots \\ V_{m_0} \end{bmatrix}$$

$$\rightarrow \widetilde{V} = \begin{bmatrix} V_1 & 0 & \dots & 0 \\ 0 & V_2 & \dots & 0 \\ \vdots & \vdots & \ddots & \vdots \\ 0 & 0 & \dots & V_{m_0} \end{bmatrix}, \tag{8.11}$$

where each V_i is $\in R^{n_i \times q}$, and V is $\in R^{N \times q}$, but $\widetilde{V}$ is $\in R^{N \times mq}$. We call this reduction a *block structure-preserving model order reduction* (BSMOR). Note that $\mathcal{V}^T \mathcal{V} = I$, i.e., each column of $\widetilde{V}$ is still linearly independent and the total column-rank is increased by a factor of the block number m.

The order-reduced state matrices are obtained by projecting $\widetilde{V}$:

$$\widetilde{\mathbf{G}} = \widetilde{V}^T \mathbf{G} \widetilde{V}, \quad \widetilde{\mathbf{C}} = \widetilde{V}^T \mathbf{C} \widetilde{V}, \quad \widetilde{\mathbf{J}} = \widetilde{V}^T \mathbf{J}. \tag{8.12}$$

Elementwise, we have

$$\widetilde{\mathbf{G}}_{i,j} = {V_i}^T \mathbf{G}_{i,j} V_j \; \widetilde{\mathbf{C}}_{i,j} = {V_i}^T \mathbf{C}_{i,j} V_j \; \widetilde{\mathbf{J}}_i = {V_i}^T \mathbf{J}_i \tag{8.13}$$

where $\widetilde{\mathbf{G}}_{i,j}$ represents the blocks at i block row and j block column in the reduced matrix $\widetilde{\mathbf{G}}$. So do $\widetilde{\mathbf{C}}_{i,j}$ and $\widetilde{\mathbf{J}}_i$. Note that such a $m \times m$ block projection preserves the structure and sparsity of the original $\mathbf{G}$, $\mathbf{C}$ matrices. For example, when projected by $\widetilde{V}$, the reduced $\widetilde{\mathbf{G}}$ matrix is

$$\widetilde{\mathcal{G}} = \begin{bmatrix} {V_1}^T \mathbf{G}_{1,1} V_1 & {V_1}^T \mathbf{G}_{1,2} V_2 & \dots & {V_1}^T \mathbf{G}_{1,m_0} V_{m_0} \\ {V_2}^T \mathbf{G}_{2,1} V_1 & {V_2}^T \mathbf{G}_{2,2} V_2 & \dots & {V_2}^T \mathbf{G}_{2,m} V_{m_0} \\ \vdots & \vdots & \ddots & \vdots \\ V_{m_0}^T \mathbf{G}_{m_0,1} V_1 & {V_{m_0}}^T \mathbf{G}_{m_0,2} V_2 & \dots & {V_{m_0}}^T \mathbf{G}_{m,m} V_{m_0} \end{bmatrix}. \tag{8.14}$$

One important observation is that if the original $\mathbf{G}$ is sparse, the resulted $\widetilde{\mathbf{G}}$ is sparse as well. In contrast, when projected by flat V using PRIMA or HiPRIME, the resulted $\hat{\mathbf{G}}$ is

$$\hat{\mathbf{G}} = \sum_{i=1}^{m} \sum_{j=1}^{m} {V_i}^T \mathbf{G}_{i,j} V_j, \tag{8.15}$$

which is dense in general. Similar sparsity holds for $\mathbf{C}$ and $\mathbf{J}$.

Because the column rank of $\mathcal{V}$ is m_0 times larger than V, the reduced model with state matrices $\widetilde{\mathbf{G}}$ and $\widetilde{\mathbf{C}}$ can approximate an additional $m_0 - 1$ poles than the reduced model with state matrices $\hat{\mathbf{G}}$ and $\hat{\mathbf{C}}$. However, the additional $m_0 - 1$ poles by $\mathcal{V}$ are not exactly matched. This projection has following limitations:

1. The system poles are determined not only by the diagonal blocks, but also determined by those off-diagonal blocks.
2. The initially provided m_0 basic blocks are not optimal for pole matching because they could have pole overlap with each other.
3. The construct of $\mathcal{V}$ needs to obtain V by reducing the entire state matrices. Its orthonormalization cost is expensive.

As discussed in Section 8.5, these problems can be solved using a triangularization-based block structure-preserving model order reduction.

8.5 TBS method

8.5.1 *Compact block formulation*

We first present a dominant-pole based clustering to generate a compact block representation of those basic blocks.

Two-level organization of basic block

We first decompose $\mathbf{G}$ and $\mathbf{C}$ in (8.9) by a two level representation

$$\mathbf{G} + s\mathbf{C} = \mathbf{Y}_0(s) + \mathbf{Y}_1(s). \tag{8.16}$$

The diagonal part $\mathbf{Y}_0(s)$ is $\mathbf{G}_0 + s\mathbf{C}_0$, where

$$\mathbf{G}_0 = \text{diag}[\mathbf{G}_{11}, ..., \mathbf{G}_{m_0 m_0}], \quad \mathbf{C}_0 = \text{diag}[\mathbf{C}_{11}, ..., \mathbf{C}_{m_0 m_0}].$$

Note that each block matrix $\mathbf{G}_{ii}$ or $\mathbf{C}_{ii}$ is symmetric positive definite (s.p.d), i.e., each basic block is passive. The off-diagonal part $(\mathbf{Y}_1)_{ij}$ is composed by the coupling block: $\mathbf{G}_{ij} + s\mathbf{C}_{ij}$ $(i \neq j)$. Its entries are usually smaller than those in basic blocks in the diagonal. Accordingly, the moment generation matrices for each basic block are

$$(\mathbf{A}_0)_i = (\mathbf{G}_{ii} + s_0\mathbf{C}_{ii})^{-1}\mathbf{C}_{ii}, \quad (\mathbf{R}_0)_i = (\mathbf{G}_{ii} + s_0\mathbf{C}_{ii})^{-1}\mathbf{J}_i.$$

As will be discussed, the two-level decomposition enables a structure-preserving model order reduction that constructs its projection matrix only from those diagonal blocks. In addition, the reduced macro-model can be efficiently analyzed in a two-level fashion.

Clustering

To obtain a more compact block representation we propose a bottom-up clustering algorithm based on the dominant poles. The system timing response for each basic block can be approximately determined by its q dominant poles, i.e., the first

q most dominant eigenvalues or poles ($\lambda_1 \leq ... \leq \lambda_q$). Poles are calculated from the eigendecomposition of the *order-reduced* moment matrix $\widetilde{\mathbf{A}} = \widetilde{\mathbf{G}}^{-1}\widetilde{\mathbf{C}}$ ($\in R^{q \times q}$). Note that when the excitation current vector is used for moment-matching the output, the size q of the reduced model with the desired accuracy can be much smaller than the size of the original model. As a result, the cost of eigendecomposition of reduced model is not high.

Precisely, for m_0 basic blocks, we calculate the first q dominant poles for each basic block by reducing it independently and finding its projection matrix V_i accordingly

$$\text{span}(V_i) = \mathcal{K}((\mathbf{A}_0)_i, (\mathbf{R}_0)_i, q) \quad i = 1, ..., m_0. \tag{8.17}$$

According to Theorem 8.2, using V_i, the reduced $\hat{\mathbf{x}}_i$ matches the first q moments of $\mathbf{x}_i$ with input $\mathbf{J}_i$. $\hat{\mathbf{x}}_i$ hence also matches the first q dominant poles of $\mathbf{x}_i$ according to Corollary 8.1.

Assume block i has $(\mathbf{G}_{ii}, \mathbf{C}_{ii}, \mathbf{J}_i)$. Its q-dominant pole set is

$$\Lambda_i = eigen[(\widetilde{\mathbf{A}}_0)_i] = \{\lambda_1 \leq ... \leq \lambda_q\}.$$

After merging block i with another block j and their interconnection $(\mathbf{G}_{ij}, \mathbf{C}_{ij})$, its q-dominant-pole set becomes

$$\Lambda_i' = eigen[(\widetilde{\mathbf{A}}_0)_i'] = \{\lambda_1' \leq ... \leq \lambda_q'\},$$

where $(\mathbf{A}_0)_i'$ is the new moment generation matrix for the merged block.

Moreover, we can define the *pole distance*. If Λ_i and Λ_j are two dominant-pole sets, $\lambda_m \in \Lambda_i$ and $\lambda_n \in \Lambda_j$, then the pole distance $d(\Lambda_i, \Lambda_j)$ is

$$d(\lambda_m, \Lambda_j) = min\{|\lambda_m - \lambda_n| : \lambda_n \in \Lambda_j\}$$
$$d(\Lambda_i, \Lambda_j) = max\{d(\lambda_m, \Lambda_j) : \lambda_m \in \Lambda_i\}.$$

The two basic blocks have a similar pole distribution and are clustered if

$$d(\Lambda_i', \Lambda_i) < \epsilon,$$

where ϵ is a small value specified by the user. More basic blocks can be merged into this cluster if they have a similar pole distribution to the cluster. On the other hand, a basic block itself is a cluster if it does not share a similar pole distribution with other blocks. The clustering obtains m clusters of basic blocks, where m is decided by the structure of P/G grids and ϵ. We call the clustera a *compact block* in this chapter. Accordingly, we can obtain a set of projection matrices: $V = [V_{1(n_1 \times q)}, \ldots, V_{m(n_m \times q)}]$, one for each compact block. The overall procedure is outlined in Algorithm 8.1.

This interconnected compact block representation reduces the complexity of the original basic block representation as fewer blocks are needed to represent the original system. Moreover, the set of the first q-dominant poles of each clustered block has a minimum overlap. Note that because the original structured power grid shows a heterogeneous structure, each region can have various RC values, and the clustering algorithm will not converge to one entire circuit. This has been verified experimentally.

Algorithm 8.1 DOMINANT-POLE CLUSTERING

$I = 0$;
/* M: unclustered blocks in m_0 */
for every i in M **do**
  input: $(\mathbf{G}_0^{(0)})_{ii}, (\mathbf{C}_0^{(0)})_{ii}, (\mathbf{B}_0^{(0)})_i$;
  find: $V_i^{(0)}$ and $(\widetilde{\mathbf{A}}_0^{(0)})_{ii}$ by PRIMA;
  /* Eigendecomposition of order reduced $\mathbf{A}$ */
  form: $\Lambda_i^{(0)} = eigen[(\widetilde{\mathbf{A}}_0^{(0)})_{ii}]$;
  for every j in M **do**
    while $(d(\Lambda_i^{(I)}, \Lambda_i^{(I+1)}) < \epsilon)$ **do**
      cluster: Merge block i and block j;
      update: Delete block i and block j in M ;
      find: $V_i^{(I+1)}$ and $(\widetilde{\mathbf{A}}_0^{(I+1)})_{ii}$ by PRIMA;
      form: $\Lambda_i^{(I+1)} = eigen[(\widetilde{\mathbf{A}}_0^{(I+1)})_{ii}]$;
      calculate: $d(\Lambda_i^{(I)}, \Lambda_i^{(I+1)})$;
    end while
    $I++$;
  end for
end for
output: $\{V_1^{(I+1)}{}_{(n_1 \times q)}, \ldots, V_m^{(I+1)}{}_{(n_m \times q)}\}$;

8.5.2 TBS model order reduction

Although clustering results in m blocks, each has the unique pole distribution; the poles of the entire grids are not only determined by those diagonal blocks. In this section, we discuss how to form the upper triangular system $(\mathcal{G}, \mathcal{C})$ that is equivalent to the original system $(\mathbf{G}, \mathbf{C})$, and show that the system poles of $(\mathcal{G}, \mathcal{C})$ are determined only by its diagonal blocks [44]. With an additional block-structured projection, the reduced blocks can match more poles than the flat projection.

Triangularization

The triangularization is based on the introduction of a replica block of $(\mathbf{G}, \mathbf{C})$, and moving those lower triangular blocks of $(\mathbf{G}_{ij}, \mathbf{C}_{ij})$ $(i < j)$ to the upper triangular parts at $(\mathcal{G}_{i,m+j}, \mathcal{C}_{i,m+j})$. The resulting *triangular system* has an *upper triangular* state matrix $\mathcal{G}$

$$\mathcal{G} = \left[\begin{array}{cccc|cccc} \mathbf{G}_{11} & \mathbf{G}_{12} & \dots & \mathbf{G}_{1m} & 0 & 0 & \dots & 0 \\ 0 & \mathbf{G}_{22} & \dots & \mathbf{G}_{2m} & \mathbf{G}_{21} & 0 & \dots & 0 \\ \vdots & \vdots & \ddots & \vdots & \vdots & \vdots & \ddots & \vdots \\ 0 & 0 & \dots & \mathbf{G}_{mm} & \mathbf{G}_{m1} & \mathbf{G}_{m2} & \dots & 0 \\ \hline & & \mathbf{0} & & & & \mathbf{G} & \end{array}\right], \tag{8.18}$$

and $\mathcal{C}$ has a similar structure to $\mathcal{G}$. The port matrix $\mathcal{B}$ and state variable x are

$$\mathcal{J} = \begin{bmatrix}\mathbf{J}_1 \ \mathbf{J}_2 \ \dots \ \mathbf{J}_m \ \mathbf{J}\end{bmatrix}^T, \ x = \begin{bmatrix}x_1 \ x_2 \ \dots \ x_m \ \mathbf{x}\end{bmatrix}^T.$$

where $\mathbf{J}$ and $\mathbf{x}$ are defined in (8.9) and (8.10).

The resulting triangular system equation is

$$(\mathcal{G} + s\mathcal{C})x(s) = \mathcal{J} \tag{8.19}$$

It is easy to verify that the solution x(s) from (8.19) is the same as $\mathbf{x}$(s) from (8.1).

Below, we prove that the new triangular system is passive.

Theorem 8.3. *The upper block triangular system ($\mathcal{G}$,$\mathcal{C}$) is passive.*

Proof: The eigenvalues of the triangular system are given by the product of determinants of diagonal blocks

$$|\mathcal{G}| = \prod_{i=1}^{m+1} |(\mathcal{G}_0)_i| = |(\mathbf{G}_0)_1|...|(\mathbf{G}_0)_m||\mathbf{G}|$$

Because each block $(\mathbf{G}_0)_i$ $(1 \le i \le m)$ and $\mathbf{G}$ are positive definite, $\mathcal{G}$ is positive definite as well. The same procedure can be used to prove that $\mathcal{C}$ is positive definite. Therefore, $\mathcal{G} + \mathcal{G}^T$ and $\mathcal{C} + \mathcal{C}^T$ are both s.p.d, and hence the triangular system is passive.

Note that directly solving (8.19) involves a similar cost of solving (8.1) as the replica block at the lower-right corner needs to be factorized first. As shown below, its benefits can be appreciated after a structure-preserving model order reduction, where the state variable of each reduced block can be solved independently with q matched poles.

mq-pole matching

After clustering in Section 3.2, we can also obtain a set of projection matrices: $\{V_1, ..., V_m, V_{m+1}\}$, where V_i $(1 \le i \le m)$ is constructed for each block. Without using orthonormalization for the replica block, V_{m+1} is obtained by

$$V_{m+1} = [V_1, ..., V_m] \quad (\in R^{N \times q}). \tag{8.20}$$

Furthermore, instead of constructing a flat projection matrix

$$V = [V_1, ..., V_m, V_{m+1}], \quad (\in R^{2N \times q}), \tag{8.21}$$

we reconstruct a block-diagonal structured projection matrix $\mathcal{V}$:

$$\mathcal{V} = diag[V_{1(n_1 \times q)}, ..., V_{m(n_m \times q)}, V_{m+1(N \times q)}], \tag{8.22}$$

with $\mathcal{V} \in R^{2N\times(m+1)q}$, $\sum_{i=1}^{m} n_i = N$. Using $\mathcal{V}$ to project $\mathcal{G}$, $\mathcal{C}$ and $\mathcal{B}$ matrices respectively, we can obtain the order-reduced state matrices

$$\widetilde{\mathcal{G}} = \mathcal{V}^T\mathcal{G}\mathcal{V}, \quad \widetilde{\mathcal{C}} = \mathcal{V}^T\mathcal{C}\mathcal{V}, \quad \widetilde{\mathcal{J}} = \mathcal{V}^T\mathcal{J},$$

In particular, the diagonal blocks in reduced $\widetilde{\mathcal{G}}$ and $\widetilde{\mathcal{C}}$ are called *reduced blocks.*

The reduced $\widetilde{\mathcal{G}}$ matrix preserves the upper block triangular structure

$$\widetilde{\mathcal{G}} = \begin{bmatrix} \widetilde{\mathcal{G}}_A & \widetilde{\mathcal{G}}_B \\ \mathbf{0} & \widetilde{\mathcal{G}}_D \end{bmatrix}, \tag{8.23}$$

where

$$\widetilde{\mathcal{G}}_A = \begin{bmatrix} \mathcal{V}_{11}^T\mathbf{G}_{11}\mathcal{V} & \mathcal{V}^T\mathbf{G}_{12}\mathcal{V}_{11} & \dots & \mathcal{V}_{11}^T\mathbf{G}_{1m}\mathcal{V}_{mm} \\ 0 & \mathcal{V}_{22}^T\mathbf{G}_{22}\mathcal{V}_{22}^T & \dots & \mathcal{V}_{22}^T\mathbf{G}_{2m}\mathcal{V}_{mm} \\ \vdots & \vdots & \ddots & \vdots \\ 0 & 0 & \dots & \mathcal{V}_{mm}^T\mathbf{G}_{mm}\mathcal{V}_{mm}^T \end{bmatrix}$$

$$\widetilde{\mathcal{G}}_B = \begin{bmatrix} 0 & 0 & \dots & 0 \\ \mathcal{V}_{11}^T\mathbf{G}_{12}\mathcal{V}_{22} & 0 & \dots & 0 \\ \vdots & \vdots & \ddots & \vdots \\ \mathcal{V}_{mm}^T\mathbf{G}_{m1}\mathcal{V}_{11} & \mathcal{V}_{mm}^T\mathbf{G}_{m2}\mathcal{V}_{22}^T & \dots & 0 \end{bmatrix}$$

$$\widetilde{\mathcal{G}}_D = \mathcal{V}_{m+1,m+1}^T\mathbf{G}\mathcal{V}_{m+1,m+1}. \tag{8.24}$$

Since BSMOR does not use triangularization, its system poles are not determined by those diagonal blocks. Therefore, its reduced macro-model does not exactly have mq poles matching (See Figure 8.1). In contrast, TBS can exactly match mq poles as discussed below.

Theorem 8.4. *For the state matrices $\mathcal{G}$ and $\mathcal{C}$ in the upper triangular block form, if there is no overlap between eigenvalues of the reduced blocks $(\widetilde{\mathbf{G}}_{ii}, \widetilde{\mathbf{C}}_{ii})$ $(\in R^{q\times q})$, i.e.,*

$$|(\widetilde{\mathbf{G}}_{00})_1 + s(\widetilde{\mathbf{C}}_{00})_1| \cup ... \cup |(\widetilde{\mathbf{G}}_{00})_m + s(\widetilde{\mathbf{C}}_{00})_m| = null, \tag{8.25}$$

the reduced system $(\widetilde{\mathcal{G}} + s\widetilde{\mathcal{C}})$ exactly matches mq poles of the original system $(\mathcal{G} + s\mathcal{C})$.

Proof: Because the original $\mathcal{G}$ and $\mathcal{C}$ are in the upper triangular form, and the projection by $\mathcal{V}$ preserves the structure, the reduced $\widetilde{\mathcal{G}}$ and $\widetilde{\mathcal{C}}$ are in the upper triangular block form as well. For an upper-triangular-block system $\widetilde{\mathcal{G}} + s\widetilde{\mathcal{C}}$, its poles (eigenvalues) are the roots of its determinant $|\widetilde{\mathcal{G}} + s\widetilde{\mathcal{C}}|$, which are determined only by the diagonal blocks

$$|\widetilde{\mathcal{G}} + s\widetilde{\mathcal{C}}| = \prod_{i=1}^{m} |\widetilde{\mathbf{G}}_{ii} + s\widetilde{\mathbf{C}}_{ii}|.$$

Note that eigenvalues of $|\widetilde{\mathcal{G}} + s\widetilde{\mathcal{C}}|$ represent the reciprocal poles of the reduced model [85]. For the reduced block $\widetilde{\mathbf{G}}_{ii} + s\widetilde{\mathbf{C}}_{ii}$ with input $\mathcal{J}_i$, its output $\widetilde{x}_i$ matches q moments and the first q domain poles of the output x_i for block $\mathbf{G}_{ii} + s\mathbf{C}_{ii}$ in the triangular system. Since the entire system consists of m compact blocks, each with unique pole distribution, the reduced model by TBS can match mq poles. Note that the redundant poles obtained from the replica block are not counted here. With more matched poles, TBS is more accurate than HiPRIME and BSMOR. This will be shown in Section 8.7.

Algorithm 8.2 Two-level analysis

1. Solve bottom level individually

$m+1$: reduced blocks

for every i in $m+1$ **do**

  (1.1) *input*: $\widetilde{\mathbf{b}}_i, (\widetilde{\mathcal{Y}}_0)_i, (\widetilde{\mathcal{Y}}_1)_i$;

  (1.2) *factor*: LU/Cholesky factor $(\widetilde{\mathcal{Y}}_0)_i$;

  $(\widetilde{\mathcal{Y}}_1)_i^{(0)} = \widetilde{\mathbf{b}}_i$;

  for every j in q **do**

    (1.3) *solve*: back-substitution $(\widetilde{\mathcal{Y}}_0)_i P_i^{(j)} = (\widetilde{\mathcal{Y}}_1)_i^{(j)}$;

  end for

end for

2. Solve top level

(2.1) *input*: $P, P^{(0)}$;

(2.2) *factor*: LU/Cholesky factor $I + P$;

(2.3) *solve*: back-substitution $(I + P)Q = P^{(0)}$;

3. Update bottom level individually

(3.1) *output*: $x = P^{(0)} - PQ$.

8.6 Two-level analysis

Because of the structure preserving, the reduced triangular system by BSMOR and TBS can be further analyzed efficiently by a two-level analysis similar to [47]. Note that a direct backward substitution-based analysis can be applied to the TBS-reduced model, as it is a triangular system. Because the two-level analysis enables the parallelized solution and can be extended to the hierarchical analysis, it is used in this chapter to obtain the solution in both frequency and time domains. As a result, the state variable of each reduced block can be solved independently with matched q poles.

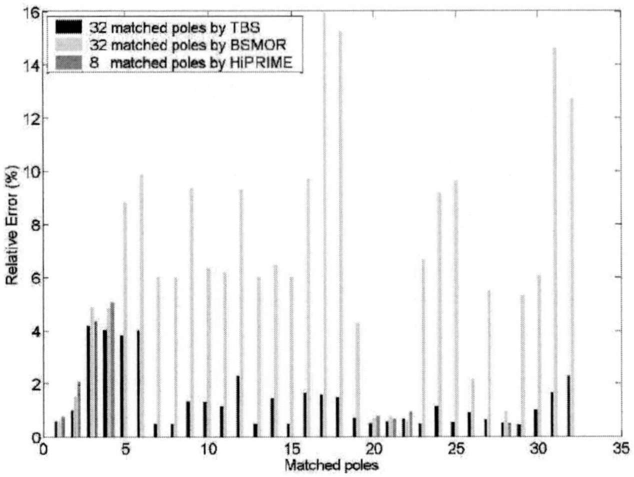

Figure 8.1 Pole matching comparison: mq poles matched by TBS and BSMOR, and q poles matched by HiPRIME. Reprinted with permission from [138] (c) 2006 ACM.

Consider the system equation for the reduced model

$$\widetilde{\mathcal{Y}}x = \widetilde{b}. \tag{8.26}$$

In the frequency domain at a frequency point s, (8.26) becomes

$$\widetilde{\mathcal{Y}} = \widetilde{\mathcal{G}} + s\widetilde{\mathcal{C}} = \widetilde{\mathcal{Y}}_0(s) + \widetilde{\mathcal{Y}}_1(s), \widetilde{b} = \widetilde{\mathcal{J}}(s),$$

and in the time domain at a time instant t with time step h, (8.26) becomes

$$\widetilde{\mathcal{Y}} = \widetilde{\mathcal{G}} + \frac{1}{h}\widetilde{\mathcal{C}} = \widetilde{\mathcal{Y}}_0(h) + \widetilde{\mathcal{Y}}_1(h), \quad \widetilde{b} = \frac{1}{h}\widetilde{\mathcal{C}}x(t-h) + \widetilde{\mathcal{J}}(t).$$

Note that the time step h can be different for each reduced block according to its dominant pole (λ_1).

The state vector $\mathbf{x}$ can be solved for each block in a fashion of two-level analysis similar to [47].

$$x = P^{(0)} - PQ, \tag{8.27}$$

where

$$P^{(0)} = (\widetilde{\mathcal{Y}}_0)^{-1}\widetilde{b}, \quad P = (\widetilde{\mathcal{Y}}_0)^{-1}\widetilde{\mathcal{Y}}_1, \quad Q = (I + P)^{-1}P^{(0)}. \tag{8.28}$$

To avoid explicit inversion, LU or Cholesky factorization needs to be applied to $\widetilde{\mathcal{Y}}_0$ and $I + (\widetilde{\mathcal{Y}}_0)^{-1}\widetilde{\mathcal{Y}}_1$. As $\widetilde{\mathcal{Y}}_0$ shows the block diagonal form, each reduced block matrix is first solved independently with LU or Cholesky factorization and substitution at

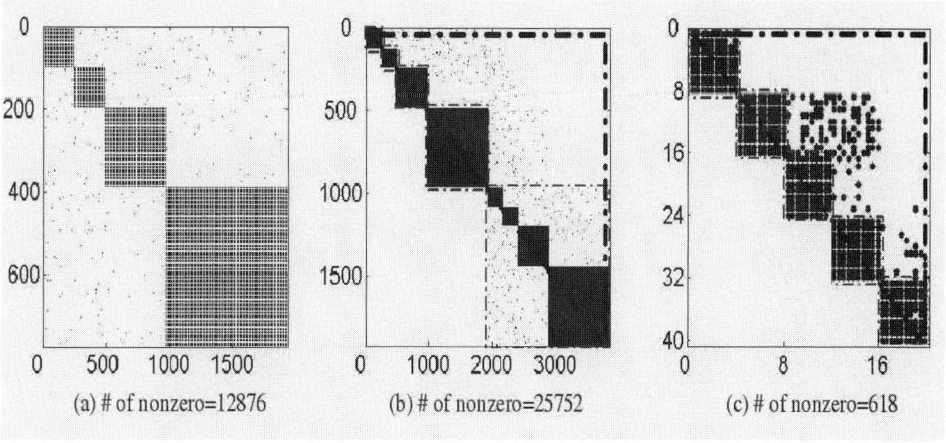

Figure 8.2 Non-zero (nz) pattern of conductance matrices: (a) original system, (b) triangular system, (c) reduced system by TBS. (a)–(c) have different dimensions, but (b)–(c) have the same triangular structure and the same diagonal block structure. Reprinted with permission from [138] (c) 2006 ACM.

the bottom level. The results from each reduced block are then used to solve the coupling block at the top level, and the final x_k of each reduced block is updated. The overall procedure is outlined in Algorithm 8.2.

8.7 Numerical examples

We implemented the BSMOR and TBS on a Linux workstation (P4 2.66GHz, 1Gb RAM). The RC mesh structures of the P/G grid are generated from realistic applications. In this section, we first verify that TBS preserves the triangular structure (sparsity) and matches mq poles, and then compare its accuracy and runtime with HiPRIME [71] and [139]. The excitation current sources (unit impulse) are explicitly considered in all MOR based methods to avoid block moment matching. The clustered block structure obtained from TBS is used as the partition for HiPRIME and [139], and the same block number is used for BSMOR but each block has the same size. The back-Euler method is used for time-domain simulation, and two-level analysis is applied for TBS, BSMOR, and [139]. In addition, TBS also considers the different time steps for different reduced blocks. In the comparison of the macro-model building and simulation time, all reduced models have similar accuracy. In the comparison of the waveform error, all MOR methods use the same number of matched moments, and macro-models for TBS and [139] have similar sizes and sparsification ratios.

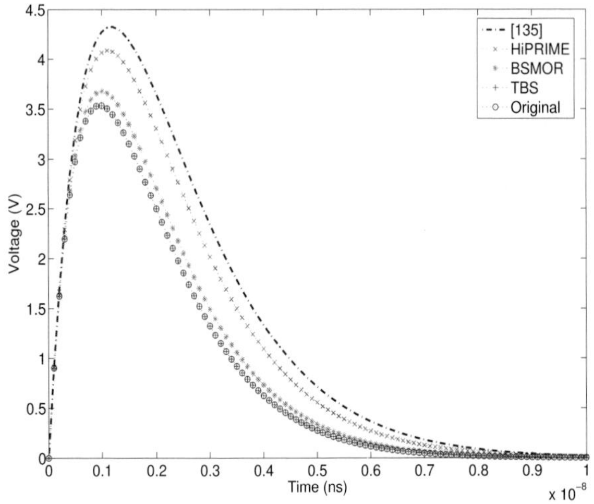

Figure 8.3 Comparison of time-domain responses between HiPRIME, BSMOR, [139], TBS and the original. TBS is identical to the original. Reprinted with permission from [138] (c) 2006 ACM.

8.7.1 A non-uniform structured RC mesh

To model the power delivery network for four layers in a 3D IC design, we use a non-uniform RC mesh (size 1M) with 32 equally sized basic blocks (each layer has eight blocks) and 32 unit-impulse current sources located at centers of basic blocks. Each basic block has a different magnitude of RC values. The number of connections between any pair of basic blocks is also different. HiPRIME, BSMOR and TBS all use $q = 8$ moments to generate the reduced model. The clustering algorithm found four clusters with 4, 4, 8, 16 basic blocks, respectively. As a result, TBS constructs a block structured projection using four blocks with the aforementioned sizes. In contrast, BSMOR constructs a block structured projection using four blocks with equal size.

Figure 8.2 shows the non-zero pattern of the conductance matrix before triangularization in Figure 8.2 (a), after triangularization in Figure 8.2 (b), and after the TBS reduction ($m = 4, q = 8$) in Figure 8.2 (c). Figure 8.2 (b) and (c) have similar non-zero patternw; this verifies that TBS preserves the triangular structure. Owing to the intrinsic sparsity by TBS, the reduced model has a 40.1% sparsification ratio. In contrast, HiPRIME generates a fully dense state matrix after the reduction and the sparsity in the reduced model by [139] is obtained by an additional LP-based sparsification.

To compare pole matching, we choose one observation port that is not at the source node. The relative errors are calculated as the magnitude difference of poles between the reduced and original models. As shown in Figure 8.1, HiPRIME can only approximate eight poles of the original model, but TBS and BSMOR can approximate 32 poles owing to increased column rank in the projection matrix.

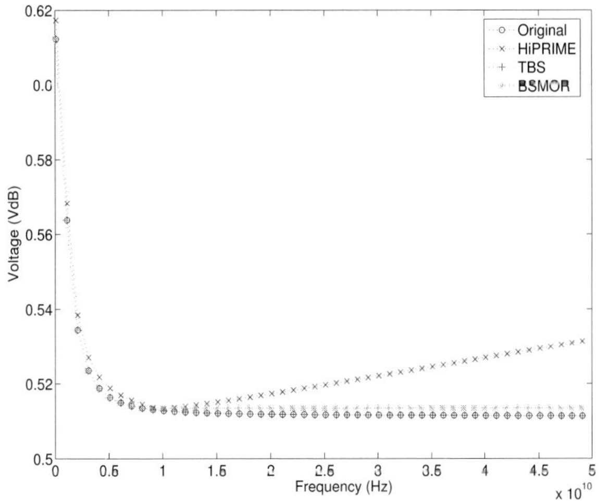

Figure 8.4 Comparison of frequency-domain responses between HiPRIME, BSMOR, TBS, and the original. TBS is identical to the original. Reprinted with permission from [138] (c) 2006 ACM.

Moreover, for poles matched by both TBS and BSMOR, TBS is about six times more accurate in average. This is because the system poles of a triangular matrix are determined by its diagonal blocks. With a structure-preserving model order reduction, the reduced triangular system by TBS can exactly match mq poles of the original system. In contrast, the reduction in BSMOR does not have the triangular structure, and hence its approximated mq poles are less accurate than those obtained by TBS.

Figure 8.3 compares the time-domain response at one port for HiPRIME, BSMOR, [139], TBS and the original under a unit-impulse input. The time-domain waveform error is counted as the relative deviation at peak voltage. The reduced model by TBS is visually identical to the original model, but HiPRIME shows up to 36% error due to much fewer matched poles, and [139] shows up to 64% error due to the sparsfication. As mentioned before, the projection matrix constructed by BSMOR uses four uniform blocks, each are the same size. As a result, it is not optimum to match poles and results in up to 23% error. Figure 8.4 further shows the frequency-domain response under an impulse input. Using the same number of moments, we observe that the reduced model by TBS is identical to the original up to 50 GHz, but the one by BSMOR or HiPRIME shows non-negligible deviation beyond 10 GHz.

8.7.2 *Scalability study*

We first study the runtime scalability of the reduced macro-model by HiPRIME, BSMOR, the method from [139] and TBS. The runtime here includes both the macro-model building time and macro-model simulation time (time-domain). All

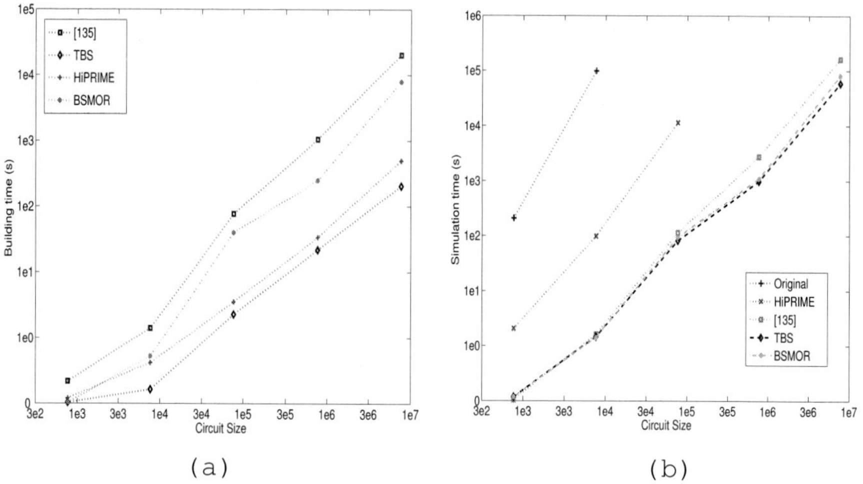

Figure 8.5 Comparison of runtime under similar accuracy. (a) macro-model building time (log scale) comparison; (b) macro-model time-domain simulation time (log scale) comparison. Reprinted with permission from [138] (c) 2006 ACM.

reduced state matrices are constructed with similar accuracy.

Figure 8.5(a) compares the macro-model building time. As [139] needs additional LP-based sparsification, it is inefficient for large sized P/G grids. For example, for a RC-mesh with size of 7.68 M, the method in [139] needs 4 hours, 42 minutes and 38 seconds to build a reduced macro-model with size 1 K and sparsity 30%, but TBS only spends 2 minutes and 8 seconds (133× speedup) to build the similar sized macro-model. Moreover, TBS also has 54× speedup over BSMOR (1 hours,45 minutes and 30 seconds) because orthonormalization is applied to each block independently in TBS. HiPRIME orthnormalizes each block independently, but its building time is still larger than TBS. This is because a higher order (4×) is required to generate a reduced model with similar accuracy as TBS. Figure 8.5(b) further compares the simulation time, where we also increase the port number when increasing the circuit size. Because HiPRIME still uses flat projection, it results in a dense macro-model that loses the structure information and cannot be analyzed hierarchically. Therefore, it becomes inefficient used for time-domain simulation. As a result, its simulation time is much larger than the other macro-models. On the other hand, BSMOR, [139] and TBS enable the two-level analysis to handle larger circuits with sizes up to 7.68 M and 1200 ports in similar runtimes. For a circuit with size $(76.8\,\text{K})^2$ and 120 ports, TBS achieves 109× runtime speedup compared with HiPRIME.

In Table 8.1, we further study the accuracy scalability of reduced macro-model by HiPRIME, BSMOR, [139] and TBS. All reduced models by MOR use the same number of moments. The standard deviation of waveform differences between the reduced and the original model is used as the measure of error. We use a higher

node (N)	port (n_p)	order (q)	TBS (m=4)	HiPRIME	BSMOR (m=4)	[139]
768	12	8	5.0310^{-7}	9.0910^{-6}	4.8710^{-6}	5.5410^{-6}
7.68K	80	40	1.8410^{-6}	2.3110^{-5}	7.9310^{-6}	1.2110^{-5}
76.8K	120	60	3.0210^{-5}	6.8210^{-4}	1.9110^{-4}	1.3110^{-2}
768K	200	100	1.2710^{-4}	9.6710^{-3}	4.2310^{-3}	6.0110^{-2}
7.68M	1200	200	3.0110^{-3}	9.9710^{-2}	5.1010^{-2}	0.11

Table 8.1 *Time-domain waveform error of reduced models by HiPRIME, BSMOR, and TBS under the same order (number of matched moments). Reprinted with permission from [138] (c) 2006 ACM.*

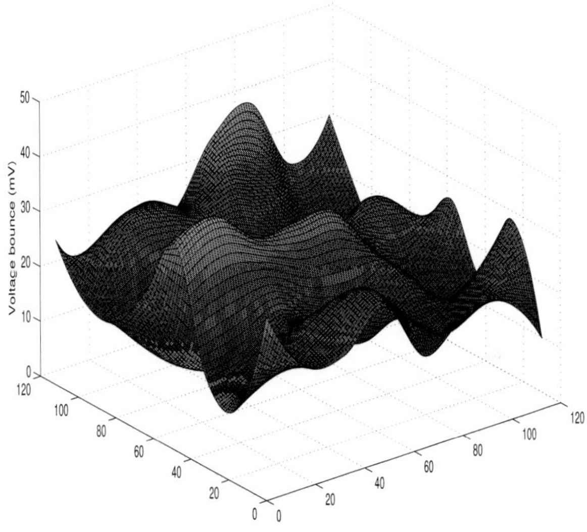

Figure 8.6 A P/G voltage bounce map without decoupling capacitor allocations. Reprinted with permission from [138] (c) 2006 ACM.

order reduced model (by 4×) as the base if the waveform of the original model is unavailable. We find that the accuracy of [139] degrades when a large sparsity ratio is needed, where LP optimization cannot preserve accuracy. On the other hand, using moment matching based projection with preserved sparsity, TBS generates a macro-model with higher accuracy. For example, it has a 38× higher accuracy than [139] when reducing a 7.68 M circuit to a 1 K macro-model with 32% sparsity. For the same circuit, TBS is 17× more accurate than BSMOR owing to the exact mq-pole matching, and is also 33× more accurate than HiPRIME owing to more matched poles.

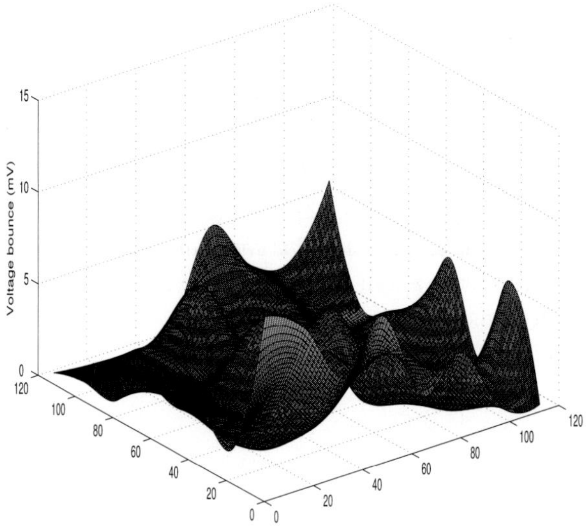

Figure 8.7 A P/G voltage bounce map with decoupling capacitors allocated at the centers of four blocks. Reprinted with permission from [138] (c) 2006 ACM.

8.7.3 Noise map for structured P/G grids

With the use of the TBS model, we can efficiently predict the locations of hotspots for large sized P/G grids, and hence, obtain the guidance to place decoupling capacitors. We apply TBS to generate a noise map of voltage bounce at a large non-uniform P/G grid (size 100 M) with 256 basic blocks of one 2D SOC design. We assume 1280 injection inputs injected uniformly in the basic blocks. The TBS method with $(m = 4, n = 64)$ is used to generate a reduced model with size 256^2 that is decomposed into four reduced blocks. We now present the time-domain noise map at four time instants under unit-impulse sources generated from the macro-model by TBS. The original model cannot finish the simulation in time domain due to the lack of memory. Figure 8.6 shows the P/G voltage bounce map at t=10 unit time. (unit time is time step h) under unit-impulse sources. Because of the charging and discharging through the interconnection between blocks the locations of hot-spots drift across the die with the maximum bounce around 45 mV. Based on this observation, we allocate four decoupling capacitors (100 pf) at the center of four blocks. As clearly shown in Figure 8.7, most hot-spots disappeared and the maximum voltage bounce was reduced to 12 mV.

8.8 Summary

In this chapter, an accurate and efficient block structure-preserving model order reduction method is introduced for large structured systems in the time-domain. By further partitioning the projection matrix into more blocks, the resulting macro-model preserves the structures such as sparsity. In contrast, the traditional model order reduction [32,34,63,71,85,113,120] all generate a dense reduced macro-model with no preserved structure. This reduction is called the BSMOR method.

Moreover, with an additional triangularization procedure, the original system is passively transformed into a form with an upper triangular block structure, where system poles are determined only by m diagonal blocks; m is decided by the nature of the structured network. With an efficient dominant-pole-based clustering and a block-structured projection, the reduced triangular system can match mq poles of original system. This reduction is called the TBS method.

Experiments show that the waveform error is reduced $33\times$ compared with the flat projection method like PRIMA and HiPRIME, and $17\times$ compared with BSMOR using user-specified partition. Moreover, with a two-level organization reduction and analysis in TBS can be performed for each block independently. Therefore, it reduces both macro-model building and simulation time. TBS is up to $54\times$ faster at building macro-models than BSMOR, and up to $109\times$ faster in simulating macro-models in the time domain than HiPRIME. In addition, as TBS preserves sparsity, it is up to $133\times$ faster at building macro-models than [139].

9 Block structure-preserving reduction for RLCK circuits

9.1 Introduction

In the chapter, we introduce another structure-preserving model order reduction method, which extends the SPRIM method [37] to more general block forms while the $2q$ moment-matching property is still preserved. The SPRIM method partitions the state matrix in the MNA (modified nodal analysis) form into natural 2×2 block matrices, i.e., conductance, capacitance, inductance, and adjacent matrices. Accordingly, the projection matrix is partitioned. As a result, SPRIM matches twice the moments of the models by using the projection matrix given by PRIMA. The reduced models also preserve the structural properties of the original models like symmetry (reciprocity). This idea has been extended to deal with more partitions by block structure-preserving model order reduction (BSMOR) [136], as shown in Chapter 8 to further exploit the regularity of the many parasitic networks. It was shown that by introducing more partitions, more poles are matched and this leads to more accurate order-reduced models [138].

However, the BSMOR method simply introduce more partitions or blocks; it does not truly preserve the circuit structures for general RLCK circuits for different input sources (voltages or currents). The reduced model does not match the $2q$ moments of the original models, as SPRIM does.

In this chapter, we first show theoretically that structure-preserving model order reduction can be applied to RLCK admittance networks, which are driven by voltage sources and requires partitioning of the original MNA circuit matrix into 2×2 block matrices. We further show that for a hybrid MNA circuit matrix where both current and voltage sources are present, $2q$ moment matching cannot be achieved. Then we proposed a general block structure-preserving reduced order modeling of interconnects (*BSPRIM*), which generalizes the SPRIM method [37] by considering more partitions. We study the matrix partitioning for both impedance matrices and admittance matrices separately and show that the reduced models will still match the $2q$ moments of the original circuits and that the circuit structures such as symmetry, sparsity can be preserved. Experiments show that with more partitions, the reduced models become more accurate.

9.2 Block structure-preserving model reduction

In this section, we review the structure-preserving projection-based MOR method SPRIM and the important results of the SPRIM method.

9.2.1 *Preliminary*

Consider a modified nodal formulation (MNA) of the RLCK circuit equation in the frequency domain:

$$\begin{aligned} \mathcal{G}x(s) + s\mathcal{C}x(s) &= \mathcal{B}i_p(s) \\ v_p(s) &= \mathcal{B}^T x(s) \end{aligned} \tag{9.1}$$

where $x(s)$ is the state variable vector, $\mathcal{G}$ and $\mathcal{C}$ ($\in R^{N\times N}$) are state matrices and $\mathcal{B}$ ($\in R^{N\times n_p}$) is

$$\mathcal{B} = [B_1^T \quad 0]^T, \tag{9.2}$$

a port incident matrix. We assume that we now have only current sources indicated by $i_p(s)$. Eliminating $x(s)$ in (9.1) gives

$$\begin{aligned} v_p(s) &= H(s)i_p(s) \\ H(s) &= \mathcal{B}^T(\mathcal{G} + s\mathcal{C})^{-1}\mathcal{B}, \end{aligned} \tag{9.3}$$

where $H(s)$ is a multiple-input multiple-output (MIMO) impedance transfer function. PRIMA finds a projection matrix V ($\in R^{N\times qn_p}$) such that its columns span the qth block Krylov subspace $\mathcal{K}(\mathcal{A}, \mathcal{R}, q)$, i.e.,

$$\text{span}V = \mathcal{K}(\mathcal{A}, \mathcal{R}, q), \tag{9.4}$$

where $\mathcal{A} = (\mathcal{G} + s_0\mathcal{C})^{-1}\mathcal{C}$, $\mathcal{R} = (\mathcal{G} + s_0\mathcal{C})^{-1}\mathcal{B}$, and s_0 is the expansion point that ensures $\mathcal{G} + s_0\mathcal{C}$ is non-singular. The resulting reduced transfer function is

$$\bar{H}(s) = \bar{\mathcal{B}}^T(\bar{\mathcal{G}} + s\bar{\mathcal{C}})^{-1}\bar{\mathcal{B}}, \tag{9.5}$$

where

$$\bar{\mathcal{G}} = V^T\mathcal{G}V, \quad \bar{\mathcal{C}} = V^T\mathcal{C}V, \quad \bar{\mathcal{B}} = V^T\mathcal{B}, \tag{9.6}$$

which has the identical expanded first qth moments of the original transfer function $H(s)$. It is called the *Grimme's projection theorem* [45]. Note that $\bar{\mathcal{G}}$, $\bar{\mathcal{C}}$ are $\in R^{qn_p\times qn_p}$, and $\bar{\mathcal{B}}$ is $\in R^{qn_p\times n_p}$.

The PRIMA-like MOR method cannot preserve the structure of the original models. This is reflected in the fact that if the impedance transfer function $H(s)$ is symmetric, the reduced transfer function $\bar{H}(s)$ is no longer symmetric. Also the reduced matrices $\bar{\mathcal{G}}$, $\bar{\mathcal{C}}$ become dense or full matrices.

9.2.2 The SPRIM Method

In [37], a structure-preserving reduced model order reduction technique, SPRIM, was proposed. The primary observation is that the impedance transfer function of RLCK circuit $H(s)$ is symmetric. By using a split projection matrix, the structure information relevant to the passive and symmetric properties of the original circuit matrices are still preserved. As a result, the reduced transfer function $\widetilde{H}(s)$ is still symmetric.

This structure preserving MOR was made possible by the observation that instead of using the Krylov subspace $\mathcal{K}(\mathcal{A}, \mathcal{R}, q)$ for the projection matrix $\widetilde{V}$, one can use any projection matrix such that the space spanned by the column in $\widetilde{V}$ contains the qth block Krylov subspace, i.e.,

$$\mathcal{K}(\mathcal{A}, \mathcal{R}, q) \subseteq \widetilde{V}. \tag{9.7}$$

The SPRIM method starts with the 2×2 structured MNA circuit matrices in the following form:

$$\mathcal{G} = \begin{bmatrix} G & A^T \\ -A & 0 \end{bmatrix}, \mathcal{C} = \begin{bmatrix} C & 0 \\ 0 & L \end{bmatrix}, \mathcal{B} = \begin{bmatrix} B_1 \\ 0 \end{bmatrix}, \tag{9.8}$$

where G ($\in R^{n_1 \times n_1}$), C ($\in R^{n_1 \times n_1}$), L ($\in R^{n_2 \times n_2}$) are conductance, capacitance and inductance matrices, and A ($\in R^{n_2 \times n_1}$) is the adjacent matrix indicating the branch current flow at the inductor. Note that $n_1 + n_2 = N$.

Therefore, a structured projection vector $\widetilde{V}$ is constructed by partitioning the projection vector V obtained from the qth PRIMA iteration,

$$V = \begin{bmatrix} V_1 \\ V_2 \end{bmatrix} \rightarrow \widetilde{V} = \begin{bmatrix} V_1 & 0 \\ 0 & V_2 \end{bmatrix}. \tag{9.9}$$

where V_1 is $\in R^{n_1 \times qn_p}$, V_2 is $\in R^{n_2 \times qn_p}$, and hence $\widetilde{V}$ is $\in R^{N \times 2qn_p}$. As a result, the number of columns in $\widetilde{V}$ is twice that in V. Accordingly the new reduced state matrices are

$$\widetilde{\mathcal{G}} = \begin{bmatrix} \widetilde{G} & \widetilde{A}^T \\ -\widetilde{A} & 0 \end{bmatrix}, \widetilde{\mathcal{C}} = \begin{bmatrix} \widetilde{C} & 0 \\ 0 & \widetilde{L} \end{bmatrix}, \tag{9.10}$$

where $\widetilde{G} = {V_1}^T G V_1$, $\widetilde{A} = {V_2}^T A V_1$, $\widetilde{C} = {V_1}^T C V_1$ and $\widetilde{L} = {V_2}^T L V_2$. Similarly, the size of $\widetilde{\mathcal{G}}$, $\widetilde{\mathcal{C}}$ ($\in R^{2qn_p \times 2qn_p}$), and $\widetilde{\mathcal{B}}$ ($\in R^{2qn_p \times n_p}$) is twice as that of $\hat{\mathcal{G}}$, $\hat{\mathcal{C}}$, and $\hat{\mathcal{B}}$ reduced by using V.

In addition to the preservation of the structure of MNA matrix, an important benefit of SPRIM is that the reduced models will match $2q$ block moments given qth block Krylov subspace $\mathcal{K}(\mathcal{A}, \mathcal{R}, q)$. The $2q$ matching property of SPRIM is mainly a result of the fact that both the original impedance and reduced impedance transfer functions are symmetric when structure is preserved. When symmetric transfer functions are reduced, $2q$ moments are matched due to the use of two Krylov subspaces.

However, the SPRIM method [37] only gives the structure-preserving MOR on the impedance matrices with current input sources. We show in the following section that admittance circuit matrices, which is driven only by voltage sources can also be reduced in a structure-preserving way.

9.3 Structure preservation for admittance transfer-function matrices

In this section, we show that for admittance transfer functions, by properly partitioning the circuit matrices and splitting the projection matrix, we can still preserve the structure of the admittance circuit matrices and achieve the $2q$ moment matching result.

9.3.1 *Circuit structure partitioning*

For an RCL circuit with voltage sources, the resulting MNA equation is written as

$$\mathcal{G}x + \mathcal{C}\frac{\mathrm{d}x}{\mathrm{d}t} = \mathcal{B}u_t(t) \tag{9.11}$$

where

$$\mathcal{G} = \begin{bmatrix} E_g^T G E_g & E_l^T & E_v^T \\ -E_l & 0 & 0 \\ -E_v & 0 & 0 \end{bmatrix}, \; \mathcal{C} = \begin{bmatrix} E_c^T C E_c & 0 & 0 \\ 0 & L & 0 \\ 0 & 0 & 0 \end{bmatrix}, \; \mathcal{B} = \begin{bmatrix} 0 \\ 0 \\ E_v^T \end{bmatrix},$$

where E_x is the incident (adjacency) submatrix for corresponding type of branches in the circuit and $u_t(t)$ is the vector of input voltage sources. The transfer admittance function becomes

$$Y(s) = \mathcal{B}^T(\mathcal{G} + s\mathcal{C})^{-1}\mathcal{B}. \tag{9.12}$$

As a result, we can still partition the circuit matrices into a 2×2 form as follows:

$$\mathcal{G} = \begin{bmatrix} G_1 & G_2^T \\ -G_2 & 0 \end{bmatrix}, \mathcal{C} = \begin{bmatrix} C_1 & 0 \\ 0 & C_2 \end{bmatrix}, \; \mathcal{B} = \begin{bmatrix} 0 \\ B_2 \end{bmatrix},$$

where G_1 and C_1, G_2 and C_2 are defined as

$$\begin{aligned} G_1 &= E_g^T G E_g, \; G_2 = \begin{bmatrix} -E_l \\ -E_v \end{bmatrix}, \\ C_1 &= E_c^T C E_c, \; C_2 = \begin{bmatrix} L & 0 \\ 0 & 0 \end{bmatrix}, \\ B_2 &= \begin{bmatrix} 0 \\ E_v^T \end{bmatrix}. \end{aligned} \tag{9.13}$$

Meanwhile, subblock matrices G_1, C_1, C_2 are positive semidefinite, i.e., they satisfy

$$G_1 \succeq 0, \; C_1 \succeq 0, \; C_2 \succ 0. \tag{9.14}$$

Notice that by comparing (9.8) and (9.13) one finds that the major difference between impedance and admittance is that the position matrices $\mathcal{B}$ are different. It turns out that this difference will not change the $2q$ moment-matching property in the structure-preserving reduction as shown in the proof of Theorem 9.1 in Section 9.7.

After getting V from qth PRIMA, we obtain $\widetilde{V}$ by partitioning V conformly according to the size of $\mathcal{G}$, $\mathcal{C}$ and $\mathcal{B}$,

$$V = \begin{bmatrix} V_1 \\ V_2 \end{bmatrix} \rightarrow \widetilde{V} = \begin{bmatrix} V_1 & 0 \\ 0 & V_2 \end{bmatrix}. \tag{9.15}$$

Here the rows of V_1, V_2 equal the rows of G_1, G_2, respectively. The new reduced matrix can be obtained by

$$\widetilde{\mathcal{G}} = \widetilde{V}^T \mathcal{G} \widetilde{V}_= \begin{bmatrix} \widetilde{G}_1 & \widetilde{G}_2^T \\ -\widetilde{G}_2 & 0 \end{bmatrix}, \widetilde{\mathcal{C}} = \widetilde{V}^T \mathcal{C} \widetilde{V} = \begin{bmatrix} \widetilde{C}_1 & 0 \\ 0 & \widetilde{C}_2 \end{bmatrix}, \; \widetilde{\mathcal{B}} = \widetilde{V}^T \mathcal{B} = \begin{bmatrix} 0 \\ \widetilde{B}_2 \end{bmatrix}.$$

The corresponding transfer function is

$$\widetilde{Y}(s) = \widetilde{\mathcal{B}}^T (\widetilde{\mathcal{G}} + s\widetilde{\mathcal{C}})^{-1} \widetilde{\mathcal{B}}. \tag{9.16}$$

It can be proved that the reduced admittance matrix $\widetilde{Y}(s)$ is still symmetric. The reciprocity property is still preserved.

9.3.2 *Re-orthonormalization of split projection matrix*

For the projection matrix $V \in R^{N \times q}$ in (9.15), its rank should be q (assuming $N > q$). After the two-way splitting operation in (9.15), the rank of $\widetilde{V}$ may however not be $2q$ and it is typically less than $2q$. The reason is that after splitting operations, some columns become linear dependent. This is also reflected in the fact that the columns in $\widetilde{V}$ is no longer orthogonal to each other, i.e.,

$$\widetilde{V}^T \widetilde{V} \neq I.$$

To mitigate this problem, we re-orthonormalize each subblock in $\widetilde{V}$ and obtain the new split projection matrix $\widetilde{T}$

$$\widetilde{T} = \begin{bmatrix} \text{orth}(V_1) & 0 \\ 0 & \text{orth}(V_2) \end{bmatrix}, \tag{9.17}$$

where, orth(X) means making the orthonormal basis for matrix X. As a result, we may end up with fewer columns in $\widetilde{T}$ than in $\widetilde{V}$, which leads to smaller sizes of the reduced models. Also, $\widetilde{T}$ become orthogonal again

$$\widetilde{T}^T \widetilde{T} = I \tag{9.18}$$

Note also that the re-orthonormalization process does not change the subspace of $\widetilde{V}$ and the moment-matching property applied to $\widetilde{V}$ is also valid for $\widetilde{T}$. But experimental results show that the re-orthonormalization can always lead to the same or more accurate reduced models.

9.3.3 Structure-preserving properties

The proposed 2×2 structure-preserving MOR method for admittance transfer function matrices has the similar properties as the SPRIM for RCL circuits containing current sources. As a result, the reduced transfer function $\widetilde{Y}(s)$ is still symmetric and the reciprocal property is preserved.

The moment-matching property can be stated as follows:

Theorem 9.1. *Choose $s_0 \in \Re$. Let $n = m_1 + m_2 + ... + m_q$ and let $\widetilde{V}$ be the matrix used in the proposed algorithm. Therefore, the first* 2q *moments in the expansions of $\boldsymbol{Y}(s)$ in* (9.12) *and the projected model $\widetilde{\boldsymbol{Y}}_{\boldsymbol{n}}(s)$ in* (9.16) *about s_0 are identical:*

$$\boldsymbol{Y}(s) = \widetilde{\boldsymbol{Y}}_{\boldsymbol{n}}(s) + o((s - s_0)^{2q}). \tag{9.19}$$

The proof sketch of the theorem is given in Section 9.7. The $2q$ moment-matching property, however, does not exist for all the RLCK circuits. For an RLCK circuit in MNA formulation, if it is driven by both current and voltage sources, the input position matrix $\mathcal{B}$ becomes

$$\mathcal{B} = [B_1^T \quad B_2^T]^T, \tag{9.20}$$

The resulting transfer function is a hybrid function matrix, which is no longer symmetric, neither are the reduced models. Therefore, we have the following result,

Corollary 9.1. *For an RLCK circuit, which is driven by both current and voltage sources, the reduced transfer function by using split projection matrix in* (9.15) *only matches the first* q *moments of the original transfer function.*

Regarding the passivity, it can be easily proved that the reduced admittance transfer function in (9.16) is positive real and thus passive.

9.4 General block structure-preserving MOR method

Structure-preserving MOR methods like SPRIM are essentially based on a 2×2 partitioning of the state matrices. To be more general than the structure-preserving algorithm for both the impedance form and the admittance form, we can make more partitions based on the algorithm mentioned before in order to get a relatively lower order model than the original order. In this section, we will introduce this general block structure-preserving model-order-reduction method, called *BSPRIM*, for both impedance and admittance transfer functions.

We show that by increasing the partition number or block number, we can match more poles using the same Krylov subspace, which leads to a more accurate reduced model as confirmed by our experimental results. BSPRIM can still match the $2q$ moments of the original transfer impedance or admittance transfer functions, which is in contrast to recent work [136]. The BSPRIM leads to localized model order reduction for subcircuits and can preserve the sparsity of the original circuit matrices.

9.4.1 Block structure-preserving MOR for impedance function matrices

For impedance transfer function matrices, to partition the original circuits into m blocks (node disjoint partitions), we need to partition the MNA circuit into $m+1$ blocks for $\mathcal{G}$, $\mathcal{C}$, and $\mathcal{B}$ respectively to preserve the structure relevant to the symmetric and moment matching properties, as shown below:

$$\mathcal{G} = \begin{bmatrix} G_1 & G_2^T \\ -G_2 & 0 \end{bmatrix} \tag{9.21}$$

where G_1, G_2 are given as

$$G_1 = \begin{bmatrix} \mathcal{G}_{1,1\,(n_1\times n)} & \cdots \mathcal{G}_{1,m-1\,(n_1\times n)} & \mathcal{G}_{1,m\,(n_1\times n)} \\ \mathcal{G}_{2,1\,(n_2\times n)} & \cdots \mathcal{G}_{2,m-1\,(n_2\times n)} & \mathcal{G}_{2,m\,(n_2\times n)} \\ \vdots & \ddots \quad \vdots & \vdots \\ \mathcal{G}_{m,1\,(n_m\times n)} & \cdots \mathcal{G}_{m,m\,(n_m\times n)} & \mathcal{G}_{m,m\,(n_m\times n)} \end{bmatrix}$$
$$G_2 = [G_{m+1,1\,(n_{m+1}\times n)}, G_{m+1,2\,(n_{m+1}\times n)}, \ldots, G_{m+1,m\,(n_{m+1}\times n)}] \tag{9.22}$$

and each block has the size n_k ($\sum_{k=1}^{m+1} n_k = N$). Then, the $\mathcal{C}$ matrix becomes

$$\mathcal{C} = \begin{bmatrix} C_1 & 0 \\ 0 & C_2 \end{bmatrix} \tag{9.23}$$

$$C_1 = \begin{bmatrix} \mathcal{C}_{1,1\,(n_1\times n)} & 0 & \cdots & 0 \\ 0 & \mathcal{C}_{2,2\,(n_2\times n)} & \cdots & 0 \\ \vdots & \vdots & \ddots & \vdots \\ 0 & 0 & \cdots & \mathcal{C}_{m,m\,(n_m\times n)} \end{bmatrix}$$
$$C_2 = \mathcal{C}_{m,m\,(n_m\times n)}. \tag{9.24}$$

The position matrices $\mathcal{B}$ are now partitioned conformly

$$\mathcal{B} = [B_1, 0] \tag{9.25}$$
$$B_1 = [B_{1\,(n_1\times n)}, B_{2\,(n_2\times n)}, \ldots B_{m\,(n_m\times n)}]^T. \tag{9.26}$$

Therefore, the reduced state matrices can be obtained as mentioned in Section 9.2.

$$\hat{\mathcal{G}}_n = \hat{V}^T\mathcal{G}\hat{V}, \quad \hat{\mathcal{C}}_n = \hat{V}^T\mathcal{C}\hat{V}, \quad \hat{\mathcal{B}}_n = \hat{V}^T\mathcal{B}. \tag{9.27}$$

To put it in more detail, every block in $\hat{\mathcal{G}}, \hat{\mathcal{C}}$ and $\hat{\mathcal{B}}$ can be expressed as

$$\hat{\mathcal{G}}_{i,j} = V_i{}^T\mathcal{G}_{i,j}V_j,\ \hat{\mathcal{C}}_{i,j} = V_i{}^T\mathcal{C}_{i,j}V_j,\ \hat{\mathcal{B}}_i = V_i{}^T\mathcal{B}_i. \tag{9.28}$$

Notice that for m-way partitioning of the original circuit, we have $(m+1)$-way partitioning of the circuit matrices.

The projection matrix $\hat{V}$, according to the partition of $\mathcal{G}$ and $\mathcal{C}$, can be obtained by $(m+1)$-way splitting of the projection matrix V obtained from the PRIMA.

$$V = \begin{bmatrix} V_{1\,(n_1 \times n)} \\ V_{2\,(n_2 \times n)} \\ \vdots \\ V_{m+1\,(n_{m+1} \times n)} \end{bmatrix}$$

$$\rightarrow \hat{V} = \begin{bmatrix} V_{1\,(n_1 \times n)} & 0 & \dots & 0 \\ 0 & V_{2\,(n_2 \times n)} & \dots & 0 \\ \vdots & \vdots & \ddots & 0 \\ 0 & 0 & \dots & V_{m+1\,(n_{m+1} \times n)} \end{bmatrix}. \tag{9.29}$$

With the reduced circuit matrix, we have the reduced-order model with the transfer function:

$$\hat{Z}_n(s) = \hat{\mathcal{B}}^T(\hat{\mathcal{G}} + s\hat{\mathcal{C}})^{-1}\hat{\mathcal{B}} \tag{9.30}$$

As a result, we have the following theorem regarding the BSPRIM method:

Theorem 9.2. *Choose* $s_0 \in \Re$. *Let* $n = m_1 + m_2 + \dots + m_q$ *and let* $\hat{V}_n$ *be the matrix used in the BSPRIM algorithm from* (9.29). *Therefore, the first* 2q *moments in the expansions of the original impedance transfer function* $\boldsymbol{Z}$ *and the reduced impedance transfer function* $\hat{\boldsymbol{Z}}_{\boldsymbol{n}}$ (9.30) *about* s_0 *are identical:*

$$\boldsymbol{Z}(s) = \hat{\boldsymbol{Z}}_{\boldsymbol{n}}(s) + o((s - s_0)^{2q}). \tag{9.31}$$

The proof sketch of this theorem is in Section 9.7. Like SPRIM, BSPRIM also preserves the passivity of the reduced models. This is reflected in the fact that both $\hat{\mathcal{G}}$ and $\hat{\mathcal{C}}$ are still positive definite after structure-preserving MOR. We then have the following result regarding the passivity:

Theorem 9.3. *The order-reduced impedance model* $\boldsymbol{Z}(s)$ *in* (9.31), *reduced by BSPRIM, is passive.*

9.4.2 Localized moment-matching concept

Although by increasing the partition number, the reduced model still matches $2q$ block moments, the reduced models actually match more poles with more partition numbers, which lead to more accurate models as confirmed by our experimental results and some earlier work [136].

The more poles matched can be explained by the concept of *localized moment matching*. We observe that the partitioned projection matrix $\hat{V}$ leads to localized projection as shown by (9.28). In other words, the block projection matrix $\hat{V}_i$ is used only for matrix blocks $\mathcal{G}_{i,x}$ and $\mathcal{G}_{x,i}$, $(x = 1, \dots m+1)$. Each structured block projection matrix $\hat{V}_i$ will lead to the localized model-order reduction for block i, which is represented by $\mathcal{G}_{x,i}$ and $\mathcal{G}_{i,x}$ matrix blocks ($x = 1, ...m$). For each block, the maximum order is much smaller than the total order of the whole circuit. As we reduce the block size (thus increasing the partition number), the matched moments (order) will get close to the maximum order of each block. Conceivably, if $\hat{V}$ reaches the same size of $\mathcal{G}$ and $\mathcal{C}$ with a sufficient partition number, all the poles of the original systems will be captured since the corresponding congruence transformation becomes similarity transformation (as $\hat{V}^T\hat{V} = I$), which preserves all the system poles.

In summary, by introducing larger partitioning number and thus more partitioned Krylov subspaces, one can obtain order-reduced models with more accuracy for each structure block by using the same Krylov subspace base vectors, or obtain the same order-reduced model (with the same accuracy) using the smaller Krylov subspace. Therefore, BSPRIM provides much more flexibilities and trade-offs between efficiency and model accuracy for reducing linear dynamic system than the traditional projection-based reduction methods such as PRIMA.

9.4.3 *Block structure-preserving MOR for admittance-function matrices*

In this case, we have $m + 1$ partitioning of the circuit matrices for m-way circuit node partitioning based on the result in Section 9.3:

$$\mathcal{G} = \begin{bmatrix} G_1 & G_2^T \\ -G_2 & 0 \end{bmatrix}, \mathcal{C} = \begin{bmatrix} C_1 & 0 \\ 0 & C_2 \end{bmatrix}, \mathcal{B} = \begin{bmatrix} 0 \\ B_2 \end{bmatrix}, \tag{9.32}$$

where G_1 and C_1 have the m-way partitioned block structure as shown in (9.22) and (9.24), respectively.

Accordingly, we split the projection matrix V obtained from PRIMA into $m + 1$ partitions to preserve the structure of $\mathcal{G}, \mathcal{C}$ and $\mathcal{B}$. When splitting V obtained from PRIMA, the number of rows of V_{m+1} should equal the row number of G_2, i.e.,

$$V = \begin{bmatrix} V_1 \\ \vdots \\ V_m \\ V_{m+1} \end{bmatrix} \rightarrow \hat{V} = \begin{bmatrix} V_1 & 0 & \dots & 0 \\ 0 & \ddots & \dots & 0 \\ \vdots & 0 & V_m & 0 \\ 0 & 0 & 0 & V_{m+1} \end{bmatrix}. \tag{9.33}$$

The corresponding $\hat{G}$, $\hat{C}$ and $\hat{B}$ are obtained as follows and the structure is preserved.

$$\hat{\mathcal{G}}_n = \hat{V}^T\mathcal{G}\hat{V}, \quad \hat{\mathcal{C}}_n = \hat{V}^T\mathcal{C}\hat{V}, \quad \hat{\mathcal{B}}_n = \hat{V}^T\mathcal{B}. \tag{9.34}$$

$$\hat{\mathcal{G}}_{i,j} = V_i{}^T\mathcal{G}_{i,j}V_j, \hat{\mathcal{C}}_{i,j} = V_i{}^T\mathcal{C}_{i,j}V_j, \hat{\mathcal{B}}_i = V_i{}^T\mathcal{B}_i. \tag{9.35}$$

It also has the following moment matching properties:

Theorem 9.4. *Choose $s_0 \in \Re$. Let $n = m_1 + m_2 + ... + m_q$ and let $\hat{V}$ be the matrix used in our proposed algorithm. Therefore, the first $2q$ moments in the expansion* *of* $\boldsymbol{Y}$ (9.12) *and the projected model* $\hat{\boldsymbol{Y}}_{\boldsymbol{n}}$ (9.16) *about* s_0 *are identical:*

$$\boldsymbol{Y}(s) = \hat{\boldsymbol{Y}}_{\boldsymbol{n}}(s) + o((s - s_0)^{2q}). \tag{9.36}$$

The proof is similar to Theorem 9.2. Similarly, the admittance-transfer function model $\boldsymbol{Y}(s)$ in (9.36), reduced by BSPRIM, is also passive.

9.5 Numerical examples

In this section, we present a number of RLC circuits to illustrate effectiveness of the proposed BSPRIM method to reduce circuit matrices formulated in both impedance and admittance forms. We compare the proposed method with PRIMA and SPRIM for different circuits and show the higher accuracy of the BSPRIM method over the two methods.

9.5.1 *Comparison results for circuits in impedance forms*

We choose a RCL interconnect circuit with current sources. Figure 9.1 shows the higher accuracy of BSPRIM over SPRIM and PRIMA. The reduced order of PRIMA is $n = 15$ and the number of blocks used in BSPRIM is $4(m = 3)$. From

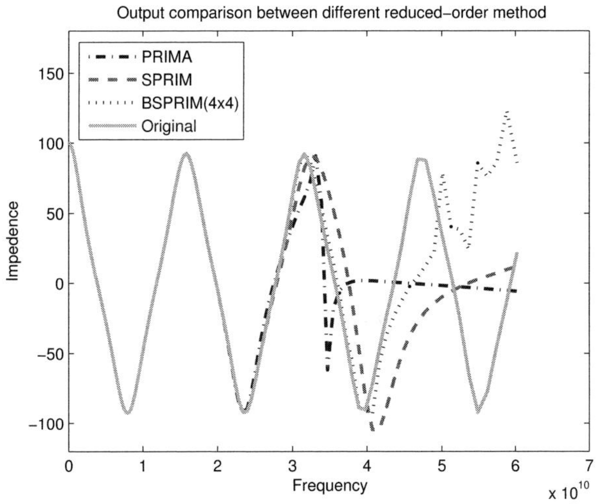

Figure 9.1 Comparison between SPRIM, PRIMA, and BSPRIM for impedance form.

Figure 9.1, we can see that SPRIM is better than PRIMA, while the 4×4 block BSPRIM has higher accuracy than SPRIM.

Figure 9.2 shows the non-zero patterns (sparsity) of $\widetilde{G}$ and $\widetilde{C}$ after reduction. While after PRIMA, the matrix G, C are almost fully dense.

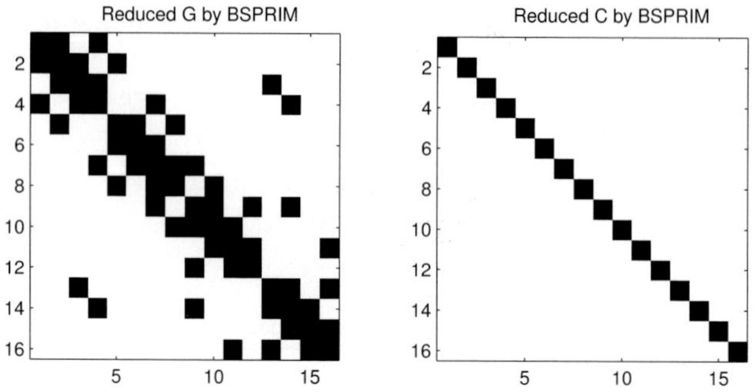

Figure 9.2 Sparsity preservation of BSPRIM.

9.5.2 Comparison results for circuits in admittance forms

In the following, we will show the experiment results of the BSPRIM algorithm for admittance form. The circuit we use consists of resistors, capacitors, inductances and voltage sources. Figure 9.3 shows the accuracy comparison of the proposed algorithm with PRIMA. The reduced order of PRIMA is $n = 15$. As a result, the reduced matrix for the proposed algorithm is $2 \times n = 30$. From Figure 9.3, the

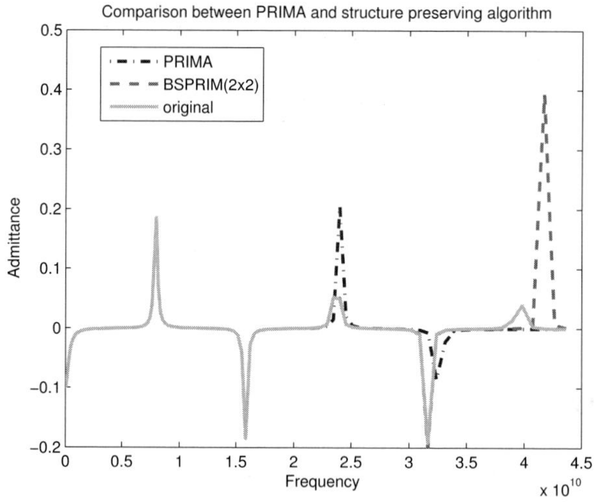

Figure 9.3 Comparison between PRIMA and structure-preserving algorithm (BSPRIM) for admittance form.

property of higher accuracy of BSPRIM for admittance form clearly can be seen.

To study the effects of re-orthonormalization process, we choose a third RLC interconnect circuit, which is driven by voltage sources; the transfer function is admittance. Figure 9.4 shows the higher accuracy of BSPRIM over the algorithms mentioned in Section 9.3 and PRIMA. For this circuit, we have a 15×15 partitioning of the original circuit. The reduced order of PRIMA is $n = 15$ and the number of blocks equals 15 ($m = 14$). From Figure 9.4, we can see that the BSPRIM method with re-orthonormalization is much more accurate than BSPRIM without re-orthonormalization. Also, after the re-orthonormalization, the dimension of the reduced-order model is reduced from 225 to 221. We have a smaller reduced model while achieving better accuracy at the same time.

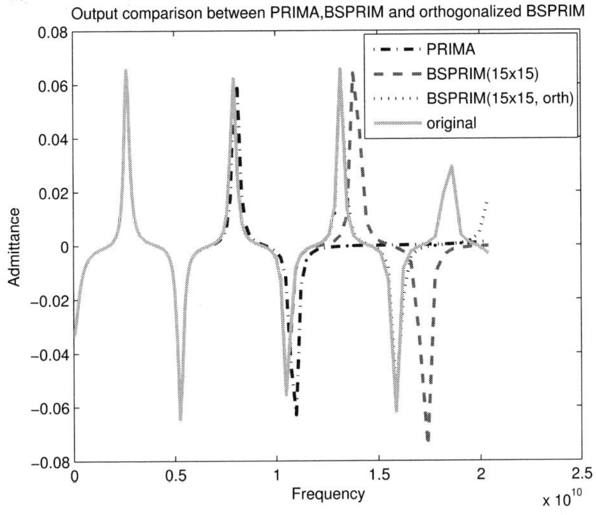

Figure 9.4 Comparison between PRIMA and BSPRIM with and without re-orthonormalization for admittance form.

9.6 Summary

In this chapter, we presented a generalized block structure-preserving reduced order interconnect macromodeling method (BSPRIM). Theoretically we show how SPRIM-like structure-preserving MOR method can be extended to deal with admittance RLC circuit matrices with $2q$ moment matching and preservation of symmetry. We also show that $2q$ moment matching cannot be achieved when the RLC circuits are driven by both current and voltage sources. We also improve SPRIM by introducing a re-orthonormalization process on the partitioned projection matrix. The new BSPRIM method can deal with more circuit partitions, which can still match the $2q$ moments and preserve circuit structure properties like symmetry, as SPRIM does. Experimental results showed that the BSPRIM outperforms SPRIM in terms of accuracy with more partitions.

9.7 Appendix

Proof of Theory 9.1: Our proof basically follows the proof in [37], if we expand in s_0, we need to show that the moment from the exact admittance function $Y(s)$ and the reduced one $\widetilde{Y}(s)$ match the $2q$ moment. That is

$$\mathcal{B}^T \mathcal{A}^j \mathcal{R}(s-s_0)^j = \widetilde{\mathcal{B}}^T \widetilde{\mathcal{A}}^j \widetilde{\mathcal{R}}(s-s_0)^j, \quad j = 0, ..., 2q-1. \tag{9.37}$$

where $\mathcal{A} = (\mathcal{G} + s_0\mathcal{C})^{-1}\mathcal{C}$, $\mathcal{R} = (\mathcal{G} + s_0\mathcal{C})^{-1}\mathcal{B}$, $\widetilde{\mathcal{A}} = (\widetilde{\mathcal{G}} + s_0\widetilde{\mathcal{C}})^{-1}\widetilde{\mathcal{C}}$, and $\widetilde{\mathcal{R}} = (\widetilde{\mathcal{G}} + s_0\widetilde{\mathcal{C}})^{-1}\widetilde{\mathcal{B}}$.

As a result, one needs to prove:

$$\mathcal{B}^T \mathcal{A}^{j_1} \widetilde{V} = \widetilde{\mathcal{B}}^T \widetilde{\mathcal{A}}^{j_1}, \; j_1 = 0, 1, ..., q. \tag{9.38}$$

and the following equation, which is the basic moment-matching theorem [45]:

$$\mathcal{A}^{j_2} \mathcal{R} = \widetilde{V} \widetilde{\mathcal{A}}^{j_2} \widetilde{\mathcal{R}}, \; j_2 = 0, 1, ..., q-1. \tag{9.39}$$

By combining (9.38) and (9.39), one can have (9.37). Hence, we mainly need to prove (9.38). Toward this line, we set

$$\mathcal{J} = \mathcal{J}^{-1} = \begin{bmatrix} I_1 & 0 \\ 0 & -I_2, \end{bmatrix} \tag{9.40}$$

where I_1, I_2 are identity matrices of the size of the diagonal blocks of $\mathcal{C}$. We have the following relations:

$$\mathcal{G}^T = \mathcal{J}^{-1}\mathcal{G}\mathcal{J}, \quad \mathcal{C} = \mathcal{J}^{-1}\mathcal{C}\mathcal{J}, \quad \mathcal{B} = -\mathcal{J}\mathcal{B}. \tag{9.41}$$

From these relations, we obtain:

$$(\mathcal{A}^T)^{j_1} = \mathcal{J}^{-1}(\mathcal{C}(\mathcal{G} + s_0\mathcal{C})^{-1})^{j_1}\mathcal{J}. \tag{9.42}$$

Since the matrix $\widetilde{\mathcal{G}}$, $\widetilde{\mathcal{C}}$ and $\widetilde{\mathcal{B}}$ have the same structure as $\mathcal{G}$, $\mathcal{C}$ and $\mathcal{B}$, the reduced-order matrices still have

$$\begin{aligned} \widetilde{\mathcal{G}}^T &= \widetilde{\mathcal{J}}^{-1}\widetilde{\mathcal{G}}\widetilde{\mathcal{J}}, \quad \widetilde{\mathcal{C}} = \widetilde{\mathcal{J}}^{-1}\widetilde{\mathcal{C}}\widetilde{\mathcal{J}}, \quad \widetilde{\mathcal{B}} = -\widetilde{\mathcal{J}}\widetilde{\mathcal{B}}, \\ (\widetilde{\mathcal{A}}^T)^{j_1} &= \widetilde{\mathcal{J}}^{-1}(\widetilde{\mathcal{C}}(\widetilde{\mathcal{G}} + s_0\widetilde{\mathcal{C}})^{-1})^{j_1}\widetilde{\mathcal{J}}. \end{aligned} \tag{9.43}$$

Combining (9.41), (9.42), and (9.39), we have

$$\begin{aligned} (\mathcal{A}^T)^{j_1}\mathcal{B} &= -\mathcal{J}^{-1}\mathcal{C}\mathcal{A}^{j_1-1}\mathcal{R} \\ &= -\mathcal{J}^{-1}\mathcal{C}\widetilde{V}\widetilde{\mathcal{A}}^{j_1-1}(\widetilde{\mathcal{G}} + s_0\widetilde{\mathcal{C}})^{-1}\widetilde{\mathcal{B}}. \end{aligned} \tag{9.44}$$

Transposing (9.44) and multiplying the right-hand side by $\widetilde{V}$, we can obtain (9.38).

$$\begin{aligned} \mathcal{B}^T\mathcal{A}^{j_1}\widetilde{V} &= \widetilde{\mathcal{B}}^T\widetilde{\mathcal{A}}^{j_1-1}\widetilde{\mathcal{J}}(\widetilde{\mathcal{C}}(\widetilde{\mathcal{G}} + s_0\widetilde{\mathcal{C}})^{-1})^T\widetilde{\mathcal{J}} \\ &= \widetilde{\mathcal{B}}^T\widetilde{\mathcal{A}}^{j_1}. \end{aligned} \tag{9.45}$$

QED.

Proof of Theory 9.2: The difference here from the proof of Theorem 9.1 is that

$$\mathcal{J}_g = \mathcal{J}_g^{-1} = \begin{bmatrix} I_1 & 0 & \dots & 0 \\ 0 & \ddots & \dots & 0 \\ \vdots & \vdots & I_m & 0 \\ 0 & \dots & 0 & -I_{m+1} \end{bmatrix}. \tag{9.46}$$

Here, I_1,I_2,...,I_m are identity matrices with the same size of V_1,V_2,...,V_m. While I_{m+1} is also an identity matrix with the same row as V_{m+1}. Then, we have

$$\begin{aligned} \mathcal{G}^T &= \mathcal{J}_g^{-1}\mathcal{G}\mathcal{J}_g, \quad \mathcal{C} = \mathcal{J}_g^{-1}\mathcal{C}\mathcal{J}_g, \quad \mathcal{B} = \mathcal{J}_g\mathcal{B}, \\ (\mathcal{A}^T)^{j_1} &= \mathcal{J}_g^{-1}(\mathcal{C}(\mathcal{G}+s_0\mathcal{C})^{-1})^{j_1}\mathcal{J}_g. \end{aligned}$$

Because of the structure-preserving property of BSPRIM, we can also get the same property of $\hat{\mathcal{G}}$, $\hat{\mathcal{C}}$, and $\hat{\mathcal{B}}$ as in (9.47).

$$\begin{aligned} \hat{\mathcal{G}}^T &= \hat{\mathcal{J}}_g^{-1}\hat{\mathcal{G}}\hat{\mathcal{J}}_g, \quad \hat{\mathcal{C}} = \hat{\mathcal{J}}_g^{-1}\hat{\mathcal{C}}\hat{\mathcal{J}}_g, \quad \hat{\mathcal{B}} = -\hat{\mathcal{J}}_g\hat{\mathcal{B}}. \\ (\hat{\mathcal{A}}^T)^{j_1} &= \hat{\mathcal{J}}_g^{-1}(\hat{\mathcal{C}}(\hat{\mathcal{G}}+s_0\hat{\mathcal{C}})^{-1})^{j_1}\hat{\mathcal{J}}_g. \end{aligned}$$

The following step is similar as the the proof of Theory 9.1. After we obtain

$$(\mathcal{A}^T)^{j_1}\mathcal{B} = \mathcal{J}_g^{-1}\mathcal{C}\widetilde{V}\widetilde{\mathcal{A}}^{j_1-1}(\widetilde{\mathcal{G}}+s_0\widetilde{\mathcal{C}})^{-1}\widetilde{\mathcal{B}}, \tag{9.47}$$

transposing and multipling the result from the right-hand side by $\hat{V}$, we will obtain

$$\begin{aligned} \mathcal{B}^T\mathcal{A}^{j_1}\hat{V} &= \hat{\mathcal{B}}^T\hat{\mathcal{A}}^{j_1-1}\hat{\mathcal{J}}_g(\hat{\mathcal{C}}(\hat{\mathcal{G}}+s_0\hat{\mathcal{C}})^{-1})^T\hat{\mathcal{J}}_g \\ &= \hat{\mathcal{B}}^T\hat{\mathcal{A}}^{j_1}. \end{aligned} \tag{9.48}$$

QED.

10 Model optimization and passivity enforcement

In this chapter, we present some general compact model optimization and passivity enforcement algorithms. Model optimization can be viewed as a fitting-based model generation process, where we fit a parameterized model against the simulated or measured data of original circuits. Model optimization methods can be applied to more general modeling applications like modeling RF and microwave circuits where it is difficult to obtain the models directly from the structures of the circuits. Instead, engineers typically model those circuits by fitting full-wave simulation or measured data. One critical issue in such applications is the preservation of some important circuit properties like passivity and reciprocity.

In this chapter, we introduce some efficient model optimization and passivity enforcement methods, as well as some reciprocity-preserving modeling methods developed in recent years.

10.1 Passivity enforcement

In this section, we present the state-space based passivity enforcement method, which is based on the method used in [21], but we will show how this method can be used in the hierarchical model order reduction framework to enforce passivity of the model order reduced admittance matrix $\tilde{\mathbf{Y}}(s)$.

Passivity is an important property of many physical systems. Brune [12] has proved that the admittance and impedance matrices of an electrical circuit consisting of an interconnection of a finite number of positive R, positive C, positive L, and transformers are passive if and only if its rational function are positive real (PR). It can be proved that the following statements are equivalent:

A network with admittance matrix function $\mathbf{Y}(s)$ is said to be positive real if and only if [5]

$$\begin{aligned}
&(1)\ \ \mathbf{Y}(s) \text{ is analytic, for } \mathrm{Re}(s) > 0,\\
&(2)\ \ \overline{\mathbf{Y}}(s) = \mathbf{Y}(\bar{s})\ , \text{ for } \mathrm{Re}(s) > 0,\\
&(3)\ \ \mathbf{Y}(s) + \mathbf{Y}(s)^H \geq 0\ , \text{ for } \mathrm{Re}(s) > 0.
\end{aligned}$$

Practically, condition (3) is difficult to check as it requires the checking of frequency responses from DC to infinite. Fortunately, there is a better way to check the positive-realness. It can be proved that the following statements are equivalent:

(a) A transfer function matrix $\mathbf{Y}(s)$ is positive real.

(b) Let $(A\ B\ C\ D)$ be a minimal controllable state space representation of $\mathbf{Y}(s)$ and $\exists K$,

$$K = K^T,\ K \geq 0, \tag{10.1}$$

such that the linear matrix inequality (LMI),

$$\left[\begin{array}{c|c} A^T K + KA & KB - C^T \\ \hline B^T K - C & -D - D^T \end{array}\right] \geq 0, \tag{10.2}$$

holds.

Constraint (10.2) actually comes from the Kalman Yakubovich Popov (KYP) lemma, which establishes important relations between state space and frequency domain objects. The KYP lemma was introduced in control theory and later used in the network synthesis [5].

If we include the term proportional to s in the transfer function, which means that we need to know what happens in infinite frequency, we can write the admittance matrix in terms of (A, B, C, D) as

$$\mathbf{Y}(s) = sY^\infty + D + C(sI - A)^{-1}B, \tag{10.3}$$

where I denotes the identity matrix with the same dimension as A. To keep the transfer function positive real, Y^∞ must satisfy

$$Y^\infty = (Y^\infty)^T,\ Y^\infty \geq 0. \tag{10.4}$$

Therefore, we can transform the problem of checking whether the admittance matrix $\mathbf{Y}(s)$ is positive real into the problem of checking whether its corresponding state space model in terms of $(A\ B\ C\ D)$ is positive semi-definite. More important is that we can use the PR criterion in terms of the state-space form to enforce the passivity of the reduced circuit matrices, as shown in the next section.

10.1.1 *State-space model representation of* $\tilde{\mathbf{Y}}(s)$

After the multi-point hierarchical model reduction, an n-port order reduced admittance matrix is generated as shown in (10.5), where each $\tilde{\mathbf{Y}}_{p,q}$ is a rational function of s. The reduction process can capture the entire dominant complex poles, which means that there are no poles in the RHP (right-hand plane) of the complex plane.

$$\tilde{\mathbf{Y}}(s) = \begin{bmatrix} \tilde{\mathbf{Y}}_{1,1} & \cdots & \tilde{\mathbf{Y}}_{1,n} \\ \vdots & \ddots & \vdots \\ \tilde{\mathbf{Y}}_{n,1} & \cdots & \tilde{\mathbf{Y}}_{n,n} \end{bmatrix}. \tag{10.5}$$

The first step we take is to transform the admittance matrix $\tilde{\mathbf{Y}}(s)$ into its state-space representation. We assume that all rational functions in the matrix share the common poles of the system. If there are private poles appearing on the leading diagonal element, we can separate them and their residues from the whole rational function after partial fraction decomposition and realize them separately.

Given a multivariable n-port network, each rational function $\tilde{\mathbf{Y}}_{p,q}$ is considered as a single-input and single-output (SISO) subsystem and mapped to its state-space representation in the controllable canonical form, which corresponds to ($A_{q,q}$, $B_{q,q}$, $C_{p,q}$, $D_{p,q}$) in the matrix of (A, B, C, D) respectively. Now we can write its state-space representation as (10.6):

$$\begin{aligned} A &= \begin{bmatrix} A_{1,1} & \cdots & 0 \\ \vdots & \ddots & \vdots \\ 0 & \cdots & A_{n,n} \end{bmatrix}, \quad B = \begin{bmatrix} B_{1,1} & \cdots & 0 \\ \vdots & \ddots & \vdots \\ 0 & \cdots & B_{n,n} \end{bmatrix}, \\ C &= \begin{bmatrix} C_{1,1} & \cdots & C_{1,n} \\ \vdots & \ddots & \vdots \\ C_{n,1} & \cdots & C_{n,n} \end{bmatrix}, \quad D = \begin{bmatrix} D_{1,1} & \cdots & D_{1,n} \\ \vdots & \ddots & \vdots \\ D_{n,1} & \cdots & D_{n,n} \end{bmatrix}. \end{aligned} \tag{10.6}$$

Also this mapping process could be viewed as n set single-input and multiple-output (SIMO) subsystems. If we choose the mth port as the input port, the mth column admittance rational function can be mapped into $(A_{m,m}\ B_{m,m}\ C_{:,m}\ D_{:,m})$.

10.1.2 Passivity-enforcement optimization

In this subsection, we briefly mention how passivity enforcement is done using a convex optimization process on the state-space representation of the admittance matrix.

Assume that we have obtained the admittance matrix of a model order reduced system $\tilde{\mathbf{Y}}(s)$ with a set of N sampling points. Let $\tilde{\mathbf{Y}}_{p,q}(s)$ denote the (p, q) entry of the transfer function $\tilde{\mathbf{Y}}(s)$. Let $\hat{\mathbf{Y}}_{p,q}(s_k)$ be the exact value of the admittance at the entry (p, q) at the kth frequency point, which can be obtained by the exact hierarchical symbolic analysis [126].

The optimization problem is to determine C, D, Y^{∞} such that a cost function is minimized with constraints on the error $(\hat{\mathbf{Y}}_{p,q} - \tilde{\mathbf{Y}}_{p,q})$. Here the constraints are on the weighted least square error, taken over N frequencies

$$\sum_{k=1}^{N} w_{k,p,q} \|\hat{\mathbf{Y}}_{p,q}(s_k) - \tilde{\mathbf{Y}}_{p,q}(s_k)\|_2^2 \leq t_{p,q}. \tag{10.7}$$

The whole optimization problem can be reformulated as the following convex programming problem:

$$\begin{aligned} \text{minimize:} \quad & t(K, C, D, Y^{\infty}), \\ \text{subject to:} \quad & (10.1), (10.2), (10.4) \\ & t \geq 0, \\ & \forall 1 \leq p, q \leq m, \ (10.7), \\ & \forall 1 \leq p, q \leq m, t_{p,q} \leq t, t \geq 0. \end{aligned} \tag{10.8}$$

where m is the port number of the network. Both the objective function and the constraints are convex functions of variables t, $t_{p,q}$, K, C, D, and Y^{∞}.

It is shown in [21] that the optimization problem in (10.7) subject to constraints in (10.2) can be transformed into a convex programming problem, which can be solved efficiently by some existing convex programming programs.

For simplicity, you can assume that $w_{p,q,k} = 1$ for all values of k, p and q. In this chapter we choose $w_{p,q,k} = 1/\|\tilde{\mathbf{Y}}_{p,q}(s_k)\|$ to normalize the relative error. Since N may be large, (10.7) will lead to a very large number of constraints. Thus a more compact form is desired.

For a matrix M, let M_p denote the pth row of M and M_q denote the qth column of M. We can write the transfer function for entry (p, q) as follows:

$$\mathbf{Y}_{p,q}(s_k) = sY_{p,q}^{\infty} + D_{p,q} + C_p(s_k I - A)^{-1}B_q. \tag{10.9}$$

Let

$$\begin{aligned} J(s_k) &= w_{p,q,k}[B_q^T(s_k I - A^T)^{-1} \ e_q^T \ se_q^T], \\ L(s_k) &= w_{p,q,k}\tilde{\mathbf{Y}}_{p,q}(s_k), \end{aligned} \tag{10.10}$$

and define

$$F_{p,q} = \begin{bmatrix} \text{Re}\{J(s)\} \\ \text{Im}\{J(s)\} \end{bmatrix}, \tag{10.11}$$

$$G_{p,q} = \begin{bmatrix} \text{Re}\{L(s)\} \\ \text{Im}\{L(s)\} \end{bmatrix}, \tag{10.12}$$

and

$$X = \begin{bmatrix} C \ D \ Y^{\infty} \end{bmatrix}. \tag{10.13}$$

we now have

$$\sum_{k=1}^{N} w_{p,q,k}\|\mathbf{Y}_{p,q}(s_k) - \tilde{\mathbf{Y}}_{p,q}(s_k)\|_2^2 = \|F_{p,q}X_p^T - G_{p,q}\|. \tag{10.14}$$

We can perform QR decomposition to the matrix $F_{p,q}$,

$$F_{p,q} = Q_{p,q}R_{p,q}, \tag{10.15}$$

where R is an upper triangular matrix and Q is an orthogonal matrix satisfying

$Q^T Q = I$. We can write

$$\|F_{p,q}X_p^T - G_{p,q}\|^2 = (F_{p,q}X_p^T - G_{p,q})^T(F_{p,q}X_p^T - G_{p,q}). \tag{10.16}$$

Let

$$E_{p,q} = (R_{p,q}X_p^T - Q_{p,q}^T G_{p,q}) \tag{10.17}$$

and

$$\delta_{p,q}^2 = G_{p,q}^T(I - Q_{p,q}Q_{p,q}^T)G_{p,q}; \tag{10.18}$$

we can rewrite (10.16) as

$$\|F_{p,q}X_p^T - G_{p,q}\|^2 = E_{p,q}^T E_{p,q} + \delta_{p,q}^2. \tag{10.19}$$

The least-square constraint (10.14) becomes

$$E_{p,q}^T E_{p,q} + \delta_{p,q}^2 \leq t_{p,q}. \tag{10.20}$$

Sometimes we need to introduce additional constraints on C, D, and Y^∞. The most common ways are probably to fix D or Y^∞ to a special value, such as zero. The fixed value must ensure that the system meets the positive-real condition.

We notice that passivity enforcement was done by the compensation-based approach proposed in [1]. But this method does not ensure the accurate matching because the compensated part may have significant impacts on the frequency range in which we are interested.

10.2 Model optimization for active circuits

10.2.1 *Introduction*

Model order reduction for passive interconnect circuits has been intensively studied in the past owing to increasing complexity of parasitic layouts of digital circuits. Many efficient algorithms have been proposed to reduce the interconnects modeled as RC/RCL/RLCK circuits, such as Krylov-subspace projection-based methods [32, 63, 85, 91, 113] and local node-reduction-based methods [3, 27, 98, 107, 108, 121].

For linear or linearized active circuits like filters, opamps etc., which dominate many analog, mixed-signal and radio-frequency (RF) circuits, fewer studies have been done to reduce those active circuits and realize the reduced circuit matrices using compact models [41]. On the other hand, the simulation of analog, mixed-signal, and RF circuits is a very time-consuming process. For instance, RF circuit simulation is an extremely time-consuming process because of very long simulation time to accommodate both the fastest and the slowest tones in the input signals [80]. As more analog RF components are manufactured on-chip using the latest digital VLSI technologies [70], parasitics, which are associated with many digital circuits, are must be considered in the integrated analog, RF circuits. This leads to increasing sizes of analog RF circuits and makes RF simulation a more challenging task. As a result, reduction and compact modeling of circuits with both passive and active circuits are important for fast mixed-signal and RF circuit design and verification.

Active circuits are different from passive circuits in several respects. First many active circuits are non-linear in nature. But they often exhibit linearity when input signals are small so that the operational points do not change significantly. Such linearized circuits are good enough for predicting many useful characteristics of the analog and RF circuits, such as gain, noise figures, bandwidth, or noise margin for early stage verification. Second, active circuits may generate energy, which means they are not passive. Hence, no passive reduction and passive enforcements are required. But the phase responses of many active circuits, such as opamps, are important as they determine the stability of the circuits caused by internal or external feedback loops [77]. But existing model order reduction approaches only match the magnitude (or real and imaginary) part of the circuit responses. Explicit matching of phase response is desired in analog circuit modeling in addition to the magnitude matching in the frequency domain. Third, active circuits typically don't have the reciprocity property [10.1]. Mathematically, a reciprocal network has a symmetric matrix, which has been exploited by many existing reduction approaches for RLCK circuits by iterative approaches. But this is not the case for general active circuits, which will make those methods less efficient. Also, the reduction process should not change the reciprocity property of the circuit during the reduction process. But this is not the case for most projection based reduction algorithms like PRIMA [91], except for the recent work [37] and the structure-preserving methods in Chapters 8 and 9. Also, how to realize general non-reciprocal (non-symmetric) order-reduced circuit matrix remains a less studied problem.

10.2.2 The new modeling method

In this section, we present a general reduction and macro-modeling technique of the linear or linearized active circuits. The new reduction process is based on the hierarchical multi-point reduction algorithm shown in Chapter 5, which allows reduction and realization of both passive and active circuits up to very wide frequency ranges [93]. The new method applies a constrained linear least-square-based optimization method to match both magnitude and phase responses of the admittance in the reduced matrix after the hierarchical reduction for modeling active circuits.

After the model generation, we can apply a general multi-port network realization process to realize any multi-port non-symmetric circuit matrices for macromodeling of non-reciprocal active circuits based on the relaxed one-port Foster's canonical form network synthesis technique, which is shown in Chapter 11. The resulting modeling algorithm can generate high-fidelity multi-port macromodels of any linear active network with easily controlled model accuracy and complexity, up to the given frequency range.

10.1 A reciprocal network is one in which the power losses are the same between any two ports regardless of the direction of propagation [127].

10.3 Optimization for magnitude and phase responses

Existing model order reduction tools typically match the admittance responses in terms of magnitudes (or real and imaginary parts). But for active circuits, phase responses are also important as they are related to the stability of the active circuits as many active circuits have internal feedback. Figure 10.1 shows the optimized admittance response of $Y_{12}(s)$ of the reduced two-port admittance matrix of the $\mu A725$ opamp circuit without considering matching phases. As can be seen, the phase part discrepancy is quite large at the low frequency range: even the magnitudes are matched perfectly. This reduced 2×2 circuit model will give incorrect phase responses, which may result in unstable or oscillating response of the whole system in time domain under some input stimulus in the given frequency range. As a result, we have to explicitly match the phase response during the optimization process shown below.

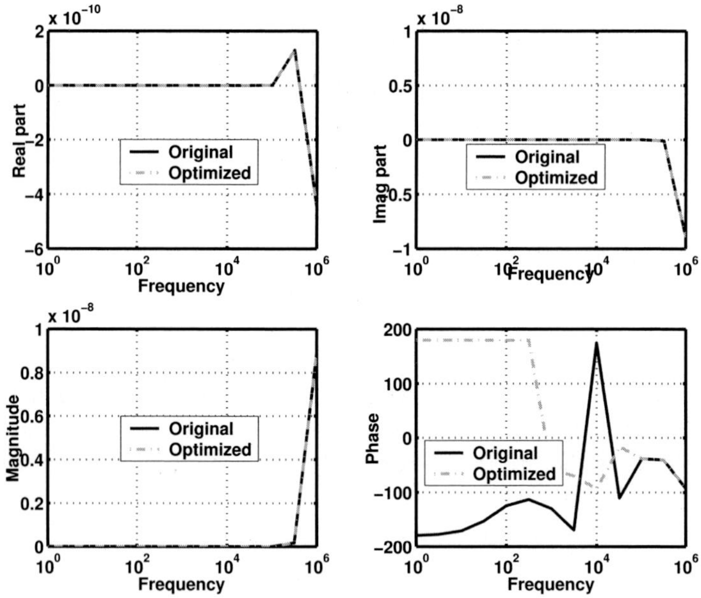

Figure 10.1 Admittance Y_{21} response of the $\mu A725$ opamp without considering phase. Reprinted with permission from [73] (c) 2005 IEEE.

10.3.1 Constrained least-square-based optimization

After the multi-point hierarchical model reduction, an n-port order reduced admittance matrix of the original circuit is generated, as shown in (10.21),

$$\hat{\mathbf{Y}}(s) = \begin{bmatrix} \hat{Y}_{1,1}(s) & \cdots & \hat{Y}_{1,n}(s) \\ \vdots & \ddots & \vdots \\ \hat{Y}_{n,1}(s) & \cdots & \hat{Y}_{n,n}(s) \end{bmatrix}, \tag{10.21}$$

where, each of the rational admittances, $\hat{Y}_{ij}$, can be represented in the partial fraction form, as shown in (10.23). The hierarchical multi-point reduction process typically finds all the dominant poles in the given frequency ranges for each admittance, but their residues may not be accurate owing to multi-point expansions. As a result, we need to adjust the residues so that admittance responses match well with the exact ones in both magnitude and phase. This can be done by the constrained least square optimization process.

Assume that we have obtained the admittance matrix of an order-reduced system $\hat{\mathbf{Y}}(s)$ with a set of T frequency sampling points. Let $\tilde{Y}_{p,q}(s_k)$ be the exact value of the admittance at the entry (p,q) at the kth frequency point, which can be obtained by the exact hierarchical symbolic analysis [126].

Let us first consider the magnitude only. The optimization problem is to determine the residues of poles such that the following least-square cost function is minimized:

$$min(\sum_{k=1}^{T} \|\hat{Y}_{p,q}(s_k) - \tilde{Y}_{p,q}(s_k)\|_2^2). \tag{10.22}$$

To format the optimization problem, we need to rewrite each rational function at the entry (p,q) of the admittance matrix in the following partial fraction form:

$$\hat{Y}(s) = s\hat{Y}_\infty + \hat{Y}_0 + \sum_{m=1}^{M} \frac{rr_m}{s - pr_m} + \sum_{n=1}^{N}(\frac{rc_n}{s - pc_n} + \frac{rc_n^*}{s - pc_n^*}), \tag{10.23}$$

where we have N-pair conjugate poles pc_n and M real poles pr_m.

For a given frequency point s_k, we define

$$A_k = [a_1^r(s_k) \cdots a_M^r(s_k)\ a_1^c(s_k) \cdots a_{2N}^c(s_k)\ 1\ s_k], \tag{10.24}$$

and

$$x = [x_1^r \cdots x_M^r\ x_1^c \cdots x_{2N}^c\ Y_0\ Y_\infty]^T. \tag{10.25}$$

For each pole lying on the real axis of s plane, we have

$$a_m^r(s_k) = \frac{1}{s_k - pr_m}, \tag{10.26}$$

and x_m^r is the residue corresponding to pole pr_m. In the case of complex poles, we have

$$a_n^c = \frac{1}{s_k - pc_n} + \frac{1}{s_k - pc_n^*}, \quad a_{n+1}^c = \frac{j}{s_k - pc_n} - \frac{j}{s_k - pc_n^*}, \tag{10.27}$$

and consequently x_n^c and x_{n+1}^c are the real and imaginary parts of the conjugate residues of the complex poles, respectively.

For T frequency points, we define

$$A_{lin} = \begin{bmatrix} \mathrm{Re}(A) \\ \mathrm{Im}(A) \end{bmatrix}, \quad Y_{lin} = \begin{bmatrix} \mathrm{Re}(\tilde{Y}) \\ \mathrm{Im}(\tilde{Y}) \end{bmatrix}. \tag{10.28}$$

We then can rewrite (10.22) as

$$\min(\|A_{lin}x - Y_{lin}\|_2^2). \tag{10.29}$$

In this way, all the variables (the real and imaginary parts of residues) are real variables and the optimization is done in the real number domain.

Now we consider the phase constraints. Phase is essentially the ratio of the real and imaginary parts of a complex number. Normally, it will be automatically matched if magnitude is matched when both real and imaginary parts are not small. But this is not the case when both of them are very small numbers. There are two aspects of the ratio: one is the sign of the ratio and the other is the value of the ratio. We first consider the sign constraint. Let us define

$$Y_D = diag(Y_{lin}), \tag{10.30}$$

where $diag(Y_{lin})$ means generating a diagonal matrix from a vector Y_{lin} and

$$D_{lin} = Y_D A_{lin}; \tag{10.31}$$

then the phase sign constraint becomes

$$D_{lin}x \geq 0, \tag{10.32}$$

Then we consider the ratio value constrain, which requires that the ratio between real and imaginary parts of the optimized one is the same as the exact one. Let's define

$$Y_I = diag\left(\begin{bmatrix} \mathrm{Im}(\tilde{Y}) \\ \mathrm{Re}(\tilde{Y}) \end{bmatrix}\right), \tag{10.33}$$

and

$$C_{lin} = I_{lin} Y_I A_{lin}, \tag{10.34}$$

where

$$I_{lin} = \begin{bmatrix} 1 & \cdots & 0 & -1 & \cdots & 0 \\ \vdots & \ddots & \vdots & \vdots & \ddots & \vdots \\ 0 & \cdots & 1 & 0 & \cdots & -1 \end{bmatrix}, \tag{10.35}$$

then the phase value constraint becomes

$$C_{lin}x = 0. \tag{10.36}$$

Finally, we have the following constrained linear least-square optimization problem:

$$min(\|A_{lin}x - Y_{lin}\|_2^2) \text{ subject to } \begin{array}{r} D_{lin}x \geq 0 \\ C_{lin}x = 0 \end{array}. \tag{10.37}$$

The resulting problem is solved by MATLAB's least square tool package in work.

10.4 Numerical examples

In this section, we present some numerical results on two examples by using the discussed optimization and passivity enforcement methods. We use SeDuMi [118] and SeDuMi Interface to solve the convex programming problem for passivity enforcement. The numerical examples of passive modeling can be found in Chapter 5. We only present the results for active circuit modeling in this section.

The first example is a folded cascode CMOS OPAMP [42]. Its small-signal model contains 122 resistors, capacitors, and voltage control current sources. We perform the multi-point hierarchical model order reduction up to 2 MHz, which extract four dominant common poles for the admittance matrix. We match the frequency up to 2 MHz during the optimization. The synthesized circuit includes 40 RLC, 2 VCVS and 2 CCCS controlled devices, which represents a 63.93% reduction ratio (63.93% circuit elements have been suppressed) for this case.

The resulting waveforms in frequency domain and comparison with the original waveforms are shown in Figure 10.2 for $Y_{12}(s)$. As one can see, the synthesized circuit matches the original circuit perfectly up to 2 MHz in all aspects of the frequency responses.

Using this CMOS opamp, we design two low-pass active filters. The first filter example is a tenth-order active Sallen–Key topology low-pass filter, shown in Figure 10.3. After the reduction and realization, there are 88 RLC elements, 2 VCVS and 2 CCCS dependent sources in the synthesized circuit compared with 636 devices in the original circuit, which represents an 85.53% reduction rate. The resulting waveforms in the frequency domain and comparison with the original waveforms are shown in Figure 10.4 for $Y_{21}(s)$. If the phase is not explicitly considered in the optimization process, the results are shown in Figure 10.5 for $Y_{21}(s)$. It can be seen that the phase part has noticeable discrepancy compared with exact response.

The filter's transfer function response is shown in Figure 10.6. Fig 10.7 shows the simulation in the time domain with different excitations. For the left figure, the input signal is a sinusoidal signal with 1 KHz frequency. The outputs of the synthesized ones are almost the same of the original circuit and the synthesized circuit. For the right figure, the excitation is also a sinusoidal function with 1 MHz frequency. The outputs are still very close.

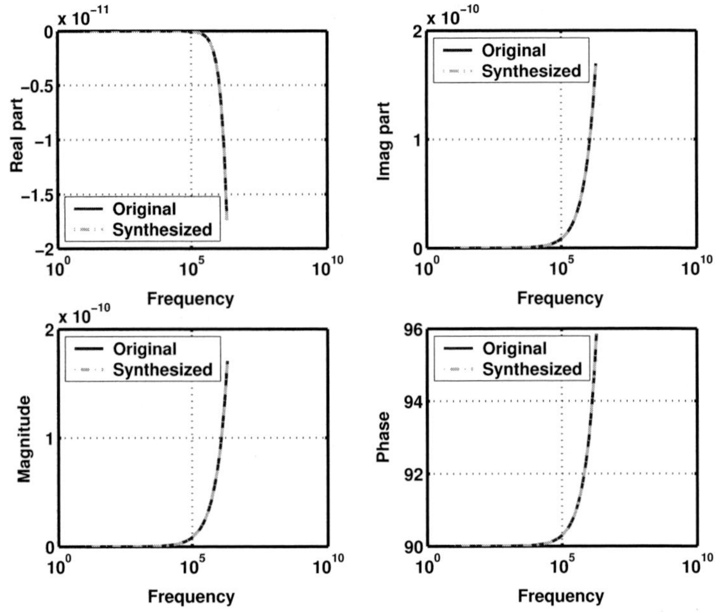

Figure 10.2 Frequency response of Y_{12} of opamp model. Reprinted with permission from [73] (c) 2005 IEEE.

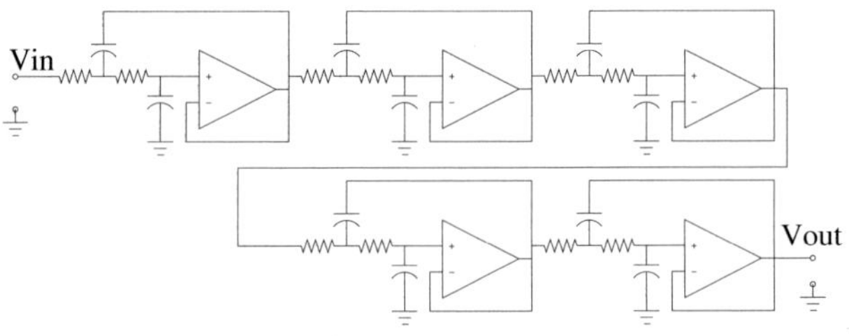

Figure 10.3 Active Sallen–Key topology low-pass filter. Reprinted with permission from [73] (c) 2005 IEEE.

The second filter is shown in Figure 10.8. This filter is a fifth-order elliptic filter using the FDNP (frequency-dependent negative resistor) technique. It contains 507 passive and active elements. The matching frequency is up to 1 MHz and we find eight dominant poles in this range. There are 56 RLC elements, 2 VCVS and 2 CCCS dependent sources in the synthesized circuit, which gives an 88.17% reduction rate.

The frequency responses of the original and synthesized circuits are shown in Figure 10.9. From this figure, we notice that the realized circuit's response is almost the same as the original system from DC to 1,MHz, but there are noticeable differences around 100 MHz. This is expected as we only match the frequency up to 1 MHz.

As a result, the original circuit can be replaced by the synthesized model if the

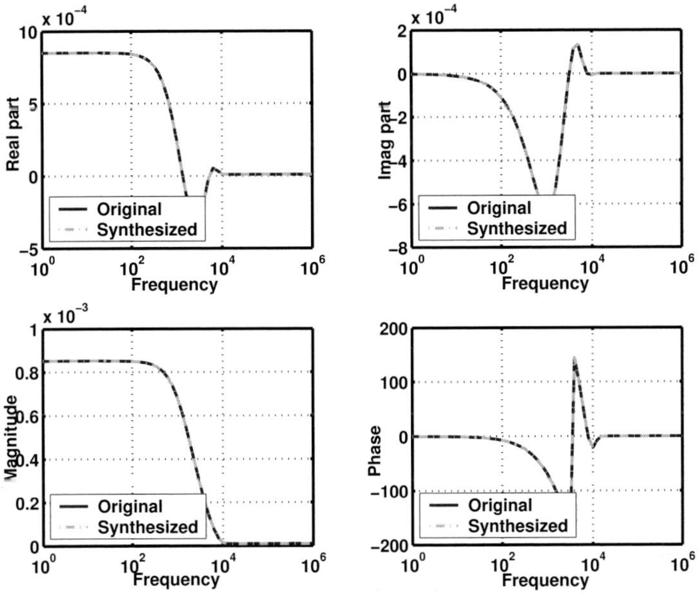

Figure 10.4 Frequency response of Y_{21} of the Sallen–Key topology low-pass filter. Reprinted with permission from [73] (c) 2005 IEEE.

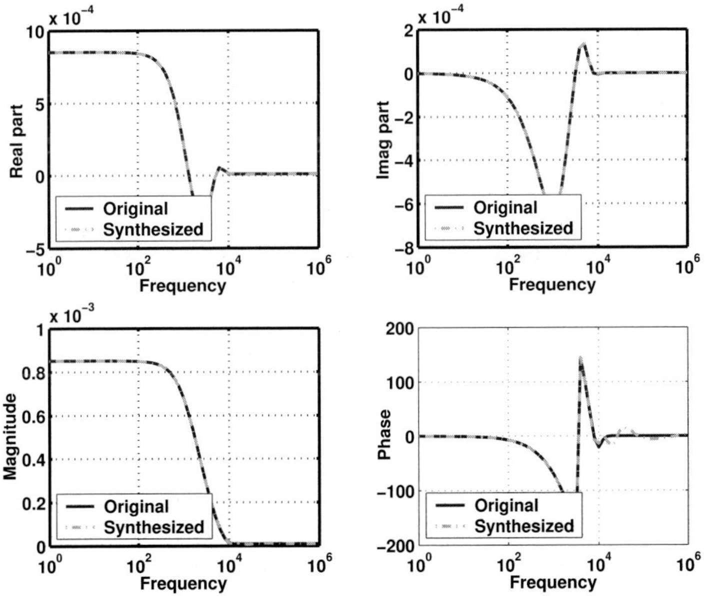

Figure 10.5 Frequency response of Y_{21} of the Sallen–Key topology low-pass filter without considering phase. Reprinted with permission from [73] (c) 2005 IEEE.

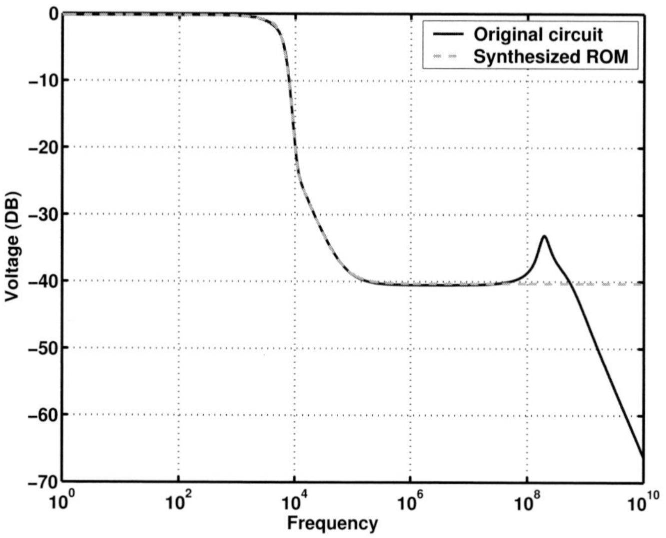

Figure 10.6 Frequency response of the transfer function of the Sallen–Key topology low-pass filter. Reprinted with permission from [73] (c) 2005 IEEE.

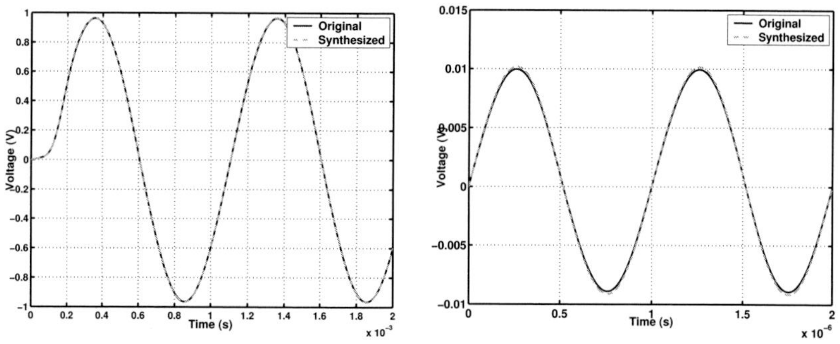

Figure 10.7 Transient response of the Sallen–Key topology low-pass filter with different excitations. Reprinted with permission from [73] (c) 2005 IEEE.

filter works in the frequency range from DC to 1 MHz (at least) or 100 MHz (at most). Over this frequency, the realized circuit will not match the original circuit well. The resulting transient waveforms are also shown in Figure 10.10. We add the same signal to the input of the original filter and the synthesized circuit to view the output waveforms. The left figure is the result from a 1Khz sinusoidal excitation. The outputs are almost same between these two filters. The right figure shows the result from a 2 MHz sinusoidal signal. The time-domain simulation is still accurate at this frequency.

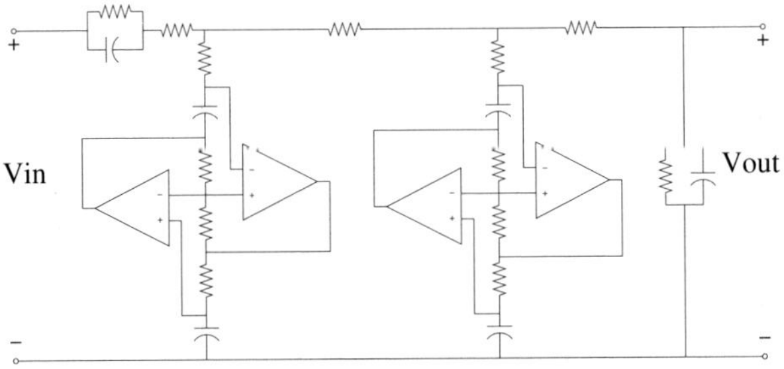

Figure 10.8 Active low-pass FDNR filter. Reprinted with permission from [73] (c) 2005 IEEE.

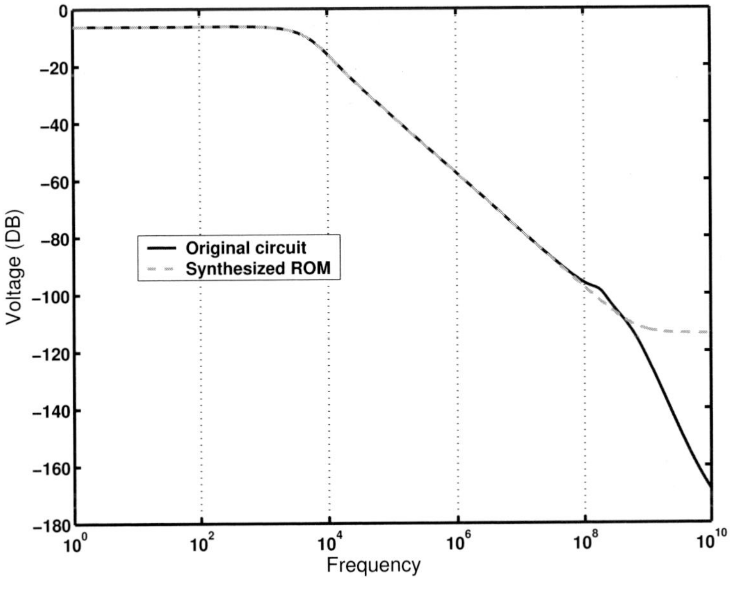

Figure 10.9 Frequency response of the transfer function of the low-pass FDNR filter. Reprinted with permission from [73] (c) 2005 IEEE.

10.5 Summary

In this chapter, we presented model optimization and passivity enforcement techniques for general passive and active circuit modeling. We first presented the convex programming or semi-definite programming based model optimization and passivity enforcement algorithm. The algorithm formulates the optimization problem into a convex programming problem with semi-definite constraints to ensure the passivity of the optimized models. This method is very general and can be applied to any fitting-based modeling process. But it suffers the scalability issue owing to expensive computation in the semi-definite programming and to the fast growing of constraints with the increasing number of terminals and poles. We then pre-

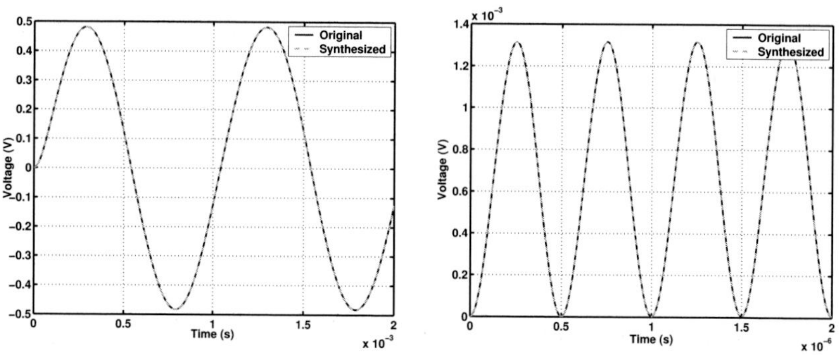

Figure 10.10 Transient response of the FDNR filter with different excitations. Reprinted with permission from [73] (c) 2005 IEEE.

sented a modeling technique for active linear analog circuits based on constrained least-square optimization. The new method explicitly considers the phase and magnitude responses of the circuits and ensures those important characteristics are well matched with that of the original circuits for active circuit modeling.

11 General multi-port circuit realization

Circuit realization deals with the issue of interfacing the reduced models with a general circuit simulation like SPICE. After the model order reduction, we will typically end up with an admittance or impedance transfer matrix $H(s) = B^T(G + sC)^{-1}B$ in the frequency domain, which can typically be rewritten as an $N \times N$ matrix,

$$H_{N\times N}(s) = \begin{bmatrix} h_{11}(s) & \cdots & h_{1n}(s) \\ \vdots & \ddots & \vdots \\ h_{N1}(s) & \cdots & h_{NN}(s) \end{bmatrix}, \tag{11.1}$$

where N is the number of ports. Each of h_{ij} can be a rational polynomial or in the partial fraction form with poles p_i and corresponding residues k_i.

There are two general ways to incorporate frequency-domain data like $H_{N\times N}(s)$ into a circuit simulator. The first method is to realize the frequency-domain data into a circuit with simple RLCM elements. The second method is by means of time-domain recursive convolution [14, 106]. Recursive convolution requires modification of the existing simulators and a new interface for getting the frequency domain data into the simulators. While realized circuits can easily be incorporated into the existing circuit simulators and are more portable among different simulators.

In this chapter, we focus on the circuit realization methods for the sake of circuit simulations. Our goal is different from the traditional circuit synthesis methods [127], where the synthesis is targeted at the physical realization of some network functions for implementing circuits, such as filters and matching networks. Instead, in our problem, we only need to build a models that can be used with circuit simulators. Such relaxed requirements can lead to significant simplification in the synthesis processes, as shown in this section.

11.1 Review of existing circuit-realization methods

Circuit realization for circuit simulation has been studied before [14, 35, 39, 98] in the computer-aided design community.

General one-port and multiple-port network realization techniques were used in the past for solving admittance synthesis problems. For example, Cauer-form-based synthesis was applied earlier by Freund and Feldmann for RC/LC synthesis by matching matrix elements of the state-matrix after the interconnect model order reduction [35, 39]. However, the matrix matching is not efficient, involving $O(N^2)$ operations, and the RC/LC template is not sufficient for general RLCM system realization. PRIME [82] is a circuit modeling and realization tool for high speed interconnect and transmission lines. It is based on Foster's form to realize all the admittance and impedance network functions. But it requires every complex pole pair to be physically realizable (every RCL element is positive), which cannot be satisfied in general and may lead to significant errors when unrealizable pole pairs are discarded or their residues are changed. In [97, 98], a template fitting approach based on convex programming was proposed. The template is borrowed from Brune's synthesis method. But this approach is expensive when high-order models are involved and it can only be applied to the one-port synthesis problem. Also the realization is an approximate process and errors will be introduced. All existing realization methods require that all the RLCM elements are physically realizable (with positive values). Therefore, the realized models are always passive.

In the following, we first present Brune's synthesis method, which can realize any one-port passive (positive-real or PR) network function with realizable RLCM elements. Brune's method is a traditional network realization method, which is computationally expensive and less suitable for our modeling purpose. But this method contain many fundamental concepts for physical network realization algorithms and it is instrumental for the reader to review those methods before we introduce a new approach to the model-oriented realization method. A comprehensive treatment of physical network realization can be found in [127].

We then present a more efficient relaxed one-port and multi-port admittance realization method based on a generalized Foster's form, which can use unrealizable (negative-valued) RLCM elements but still preserve the passivity. The new method is an exact realization method without introduction of any error and also overcomes the over-constrained synthesis conditions of the PRIME method.

11.1.1 One-port realization framework by RLCM elements

In this section, we first review the conditions for positive real functions and then represent the general realization framework for one-port network by using RLCM elements.

Positive real function revisited

The RLCM immittances function $Z(s)$ of a one-port network is positive real (passive) only if [127]:

1. $Z(s)$ is a real rational function of s,
2. $\mathrm{Re}Z(j\omega) \geq 0$ for all real ω,
3. All poles of $Z(s)$ are in the closed LPH; all $j\omega$-axis poles are simple, with positive real residues.

It is shown in [127] that these conditions are also sufficient for a one-port network to be physically realizable. By using the traditional one-port network realization technique, the positive-real property will guarantee that all the RLC elements are positive (thus physically realizable).

General one-port network realization steps

The basic idea of physical one-port network realization is to break the impedance or admittance network into smaller realizable networks, which can be implemented by simple L, C, and R elements connected in series or parallel.

There are two techniques used in the synthesis of one-port networks: (1) pole removal and (2) constant removal. The first operation realizes a series reactance branch when used on an impedance function $Z(s)$ and a shunt susceptance branch when performed on an admittance function $Y(s)$. The second operation realizes a series resistor from a $Z(s)$ and a shunt conductance from a $Y(s)$ function.

Notice that for the pole-removal operations, we require that all the poles are in the $j\omega$-axis. The reason is that poles or pole pairs in the $j\omega$-axis can easily be realized by LC elements in series or in parallel.

For instance, a network function whose impedance and admittance forms are as follows:

$$Z(s) = s + \frac{1}{s} + Z_1(s) = \frac{s^2+1}{s} + Z_1(s). \tag{11.2}$$

$$Y(s) = \frac{s}{s^2+1} + Y_1(s). \tag{11.3}$$

Note that $Z(s)$ has at least two poles, $s = 0$ and $s = \infty$. So it can be realized by a series branch with LC in series, as shown in Figure 11.1(a). Similarly, $Y(s)$ has at least two poles at $s = -j$ and $s = j$, which can be realized by a shunt LC branch in series, as shown in Figure 11.1(b).

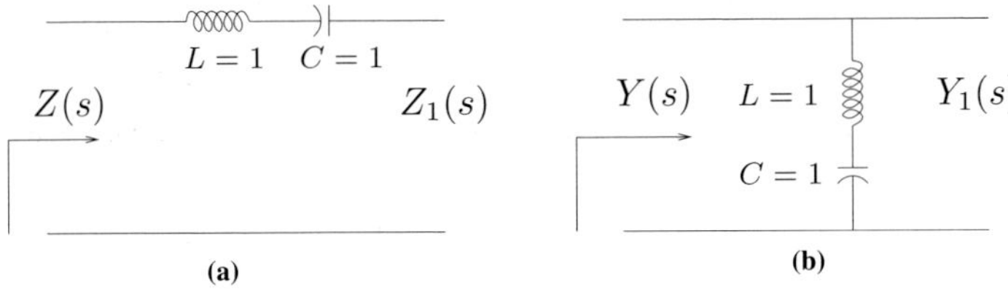

Figure 11.1 Realization of $Z(s)$ in (11.2) and $Y(s)$ in (11.3).

Similarly a network whose impedance is the admittance in (11.3) is shown below:

$$Z(s) = \frac{s}{s^2+1} + Z_1(s). \tag{11.4}$$

$$Y(s) = s + \frac{1}{s} = \frac{s^2+1}{s} + Y_1(s). \tag{11.5}$$

Now $Z(s)$ has at least two poles, at $s=-j$ and $s=j$, which can be realized by a series LC branch in parallel, as shown in Figure 11.2(a). Similarly, $Y(s)$ has at least two poles, $s=0$ and $s=\infty$. Hence, it can be realized by a shunt branch LC with in parallel, as shown in Figure 11.2(b).

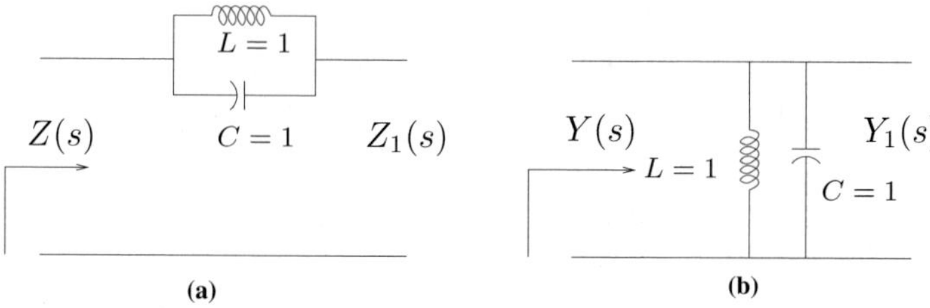

Figure 11.2 Realization of $Z(s)$ in (11.4) and $Y(s)$ in (11.5).

In the following, we use the example to illustrate the synthesis process [127]. Consider a positive-real impedance function

$$Z(s)=\frac{2s^3+4s^2+2s+3}{s^3+3s^2+4s+1}. \tag{11.6}$$

Since the degrees of the numerator and the denominator are the same, there is no pole or zero at $s=\infty$. For $s=0$, $Z(0)=3$. Hence there is no pole or zero at $s=0$ either. We notice that s^2+1 is the common factor in the numerator. Hence $Y(s)=1/Z(s)$ has poles at j and $-j$. These poles can be removed by the partial-fraction expansion:

$$Y(s)=\frac{k_1 s}{s^2+1}+Y_1(s),$$

where

$$k_1=\left[\frac{s^2+1}{s}Y(s)\right]_{s=j}=1,$$

and

$$Y_1(s)=\frac{s+1}{3s+3}=\frac{s}{6s+9}+\frac{1}{3}.$$

So $Y_1(s)$ can be realized by a shunt resistor with a RC branch. The realized circuit for $Z(s)$ in (11.6) is shown in Figure 11.3.

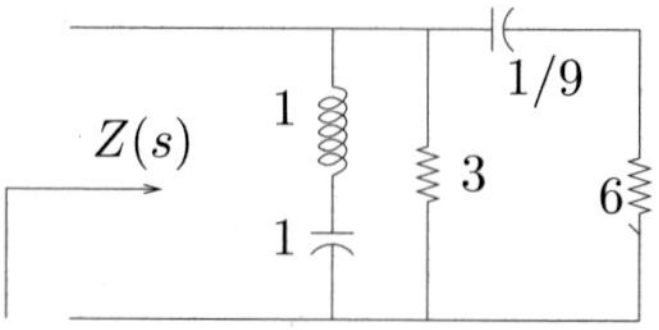

Figure 11.3 Realization of $Z(s)$ in (11.6).

We summarize the one-port network realization process for non-minimum functions as follows [127]:

1. Check $Z(s)$ for $j\omega$-axis poles. If there are any, remove them, thereby realizing a series reactance branch of the circuit.
2. Check remainder $Z_1(s)$ for $j\omega$-axis zeros. If there are any, remove the corresponding poles from $Y_1(s) = 1/Z_1(s)$. Repeat steps 1 and 2 until there are no $j\omega$-axis poles and zeros in the remainder $Z(s)$ or $Y(s)$.
3. Find the minimum value which the real part of the remainder $Z(s)$ or $Y(s)$ assuming along the $j\omega$ axis. Remove this value as a series resistor (or shunt conductance). Return step 1.

It was shown in [127] for a positive-real function, the synthesis process will always generate positive R, L, and C values to ensure the physical realizable of the synthesized circuits.

For practical interconnect and network functions extracted or obtained from full-wave simulation or measurement, most of the poles of network functions are not on the $j\omega$-axis as shown in [20]. As a result, the realization process will not work for those network functions. In the sequel, we introduce Brune's method, which can realize any positive-real function.

11.1.2 *One-port realization by Brune's method*

The driving point synthesis for RLCM circuits was studied earlier by Brune in his significant paper [12], where he pointed out positive-real impedance $Z(s)$ can be realized by a passive multiple-stage ladder network. To understand Brune's method, we first introduce the concept of the *minimum function*.

A minimum function is a function that cannot be synthesized by pole and constant removal as mentioned in the previous subsection. The real part of this function will reach 0 at a finite value ω_1 along the $j\omega$-axis, as shown in Figure 11.4 and indicated by $Z_1(\omega)$ in dotted line.

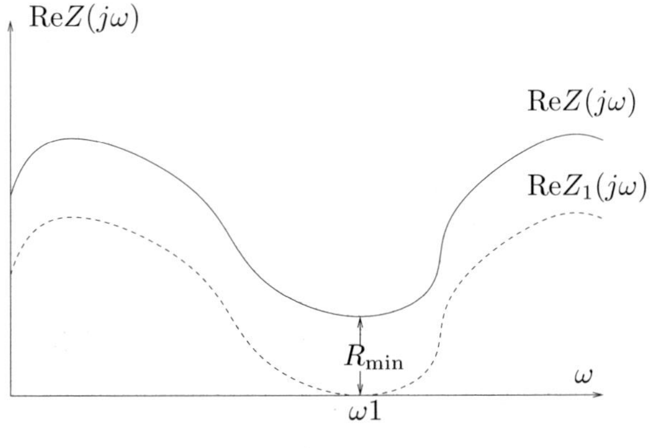

Figure 11.4 Real-part responses of $Z(s)$ and the remainder $Z_1(s) = Z(s) - R_{\min}$.

Practically, a given network function, $Z(s)$, may be a constant away from the minimum function as shown in Figure 11.4. After we remove the constant, which will be realized by a series resistor, the remainder function $Z_1(s)$ becomes a minimum function.

The idea of Brune's method is to *create* a pole in the remainder function explicitly. To do this, we notice that the real part of $Z(\omega)$ is zero at $j\omega_1$. So we have

$$Z_1(j\omega_1) = jX_1. \tag{11.7}$$

We try to extract a series element $Z_s(s)$ such that the remainder $Z_2(s)$ is zero at $j\omega_1$,

$$Z_s(j\omega_1) = Z_1(j\omega_1) - Z_2(j\omega_1) = jX_1. \tag{11.8}$$

Assume that $X_1 < 0$ and the conclusion is also applied to $X_1 = 0$ and $X_1 > 0$.

So to realize $Z_s(s)$, we can use a capacitor C_1, such that $C_1 = 1/(\omega_1|X_1|) > 0$. However, for the remainder $Z_2(s)$

$$Z_2(s) = Z_1(s) - 1/(sC_1),$$

will then become non-positive-real as it has a pole at $s = 0$ with negative residue $-1/C_1$. Note that $Z_1(s)$ does not have pole in $j\omega$ axis.

Hence, instead of using a positive capacitor, we use *negative* inductor directly to realize $Z_s(s)$ and we show that we can transform the negative inductor L_1 into physical realization transformers later.. So we have

$$Z_2(s) = Z_1(s) - sL_1 = Z_1(s) + s|L_1|, \tag{11.9}$$

which is the sum of the two positive-real functions and thus positive real itself. Since it has a zero at $s_1 = j\omega_1$ and a pole at $s = \infty$, we can now perform the normal pole removal as discussed above. First, we remove the poles at $j\omega_1$ (also at $-j\omega_1$) for $Y_2(s) = 1/Z_2(s)$

$$Y_2(s) = \frac{k_1 s}{s^2 + \omega_1^2} + Y_3(s), \tag{11.10}$$

where

$$k_1 = \left[\frac{s^2 + \omega_1^2}{s} Y_s(s)\right]_{s=j\omega_1}. \tag{11.11}$$

As a result, we have a shunt RL branch with $L_2 = 1/k_1$ and $C_2 = k/\omega_1^2$. Then we have

$$Y_3(s) = Y_2(s) - \frac{s/L_2}{s^2 + \omega_1^2}, \tag{11.12}$$

which is positive real as it comes from (11.10).

Note that $Z_2(s)$ has a pole at $s = \infty$, which means that $Y_3(s)$ has a zero at $s = \infty$. Hence $Z_3(s)$ has a pole at $s = \infty$. This pole can be removed by extracting a positive inductor L_3, and the remainder becomes

$$Z_{rem} = Z_3(s) - sL_3, \tag{11.13}$$

which is positive real too. It can be shown that L_3 can be computed as [127]

$$L_3 = \frac{-L_1 L_2}{L_1 + L_2} = \frac{|L_1| L_2}{-|L_1| + L_2}. \tag{11.14}$$

The Brune's realization cycle and the realized circuit are illustrated in Figure 11.5

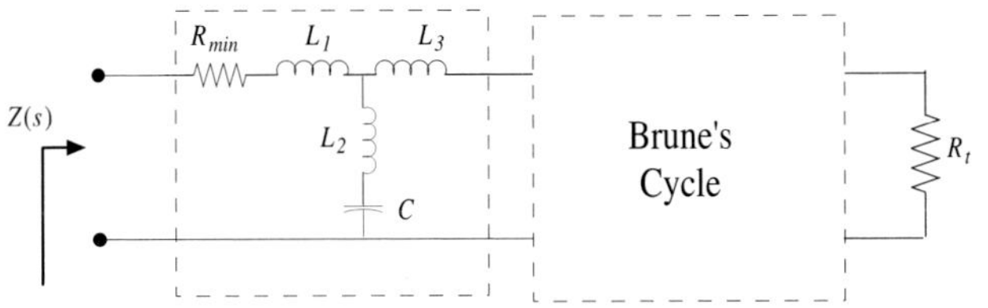

Figure 11.5 Brune's driving point synthesis by multiple-stage RLCM ladders (Brune's cycle).

Note that these synthesis steps stay at the mathematical concept level. Next, we present a practical algorithm to implement these synthesis steps [137]. The outline of the algorithm is illustrated in Figure 11.6

```
ONE PORT BRUNE SYNTHESIS(Z0(s), L)
While (i ≤ L and Zf is positive real) {
  Do jω-axis pole/zero check/remove {
    Z1(s) = Check/remove jω-axis pole (Z0(s));
    Z2(s) = Check/remove jω-axis zero (Z1(s));
  }
  B(s) = Check/remove jω-axis minimum resistance (Z2(s));
  Brune's RLCM ladder synthesis (B(s));
  Z0(s) = Zf(s) = Remainder impedance;
  i++;
}
Calculate termination resistor Rt by Zf(s);
```

Figure 11.6 Brune's multiple level ladder macromodel synthesis.

We will then briefly illustrate this procedure by studying an example with the following driving point impedance:

$$Z(s) = \frac{8s^2 + s + 4}{24s^3 + 11s^2 + 20s + 1}.$$

To obtain a minimum function, it is first necessary to remove all possible poles and zeros of $Z(s)$ from $j\omega$-axis to obtain a remainder for further minimum impedance removing. According to Brune's synthesis rule discussed above, (i) removing a zero results in a series resonant LC element in parallel with the remainder

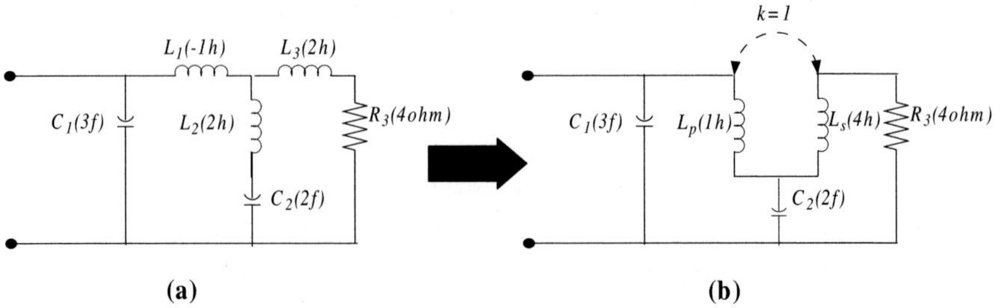

Figure 11.7 Example of Brune's synthesis with passivity-preserved transformation: the non-passive T circuit is transformed to a passive coupled-inductor circuit.

impedance; and (ii) removing a pole results in a parallel resonant LC element in series with the remainder impedance. All these operations preserve the positive-real property. In this example, there is a zero for $s \to \infty$. After removing this zero we add $C_1 = \frac{Y(s)}{s}|_{s\to\infty} = 3F$ according to rule (i). We then obtain a remainder $Z_1(s)$ for further minimum resistance removing. This minimum impedance is obtained by numerically solving the root ω_1 of

$$\frac{\mathrm{d}(\mathrm{Re}Z_1(s))}{\mathrm{d}\omega} = 0. \tag{11.15}$$

In this case, we find $\omega_1 = 0.5$ and its corresponding minimum resistance: $R = \mathrm{Re}Z_1(s) = 0.0$. Hence we have a minimum function: $B(s) = Z_1(s) - R = Z_1(s)$. According to Brune's synthesis rule (iii), by removing this resistance R, the $B(s)$ function becomes pure imaginary impedance at $j\omega_1$: $Z_1(s)|_{s=0.5j} = -0.5j$, and it can be synthesized by a negative inductance $L_1 = \mathrm{Im}Z_1(s)/\omega_1 = -1H$. Although we see a negative inductance here, we show in the following that by transforming the resultant non-passive T circuit into a pair of coupled inductors, we can still preserve the passivity. A general proof of this property can be found in [127].

Now the remainder becomes:

$$Z_2(s) = B(s) - sL_s = \frac{8s^3 + 16s^2 + 2s + 4}{8s^2 + 8s + 1},$$

and it has a zero at $j\omega_1 = 0.5$. Hence by further removing this zero of $Z_2(s)$, we can synthesize a series $L_2 = 2H$ and $C_2 = 2F$. Finally, we obtain a remainder $Z_f(s)$ that cannot be further reduced: $Z_3(s) = 2s + 4$. Hence we can realize it by a inductance $L_3 = 2H$ and a termination resistance $R_t = 4\Omega$ as shown in Figure 11.7 (a). We further apply the passive transformation as shown in Figure 11.7 (b), and obtain passive parameters of a pair of coupled inductors: $L_p = L_1 + L_2 = 1H$, $L_s = L_2 + L_3 = 4H$, and $M = L_2 = 2H$. Note that the coupling between L_p and L_s is perfect $(\frac{M}{\sqrt{LpLs}} = 1)$.

Because we have an order-reduced driving point impedance function, it enables us to practically synthesize a low-order driving point impedance macromodel. As shown in Figure 11.6, we control the synthesized Brune's ladder stage by a specified number L, where we terminate the procedure and add a termination resistor R_t:

$$R_t = Z_f(s)|_{s\to 0}, \tag{11.16}$$

where $Z_f(s)$ is the remaining impedance in the final synthesis step, and this termination resistor R_t takes accounts of the remaining DC impedance. We note that usually after several stage extractions, the extracted macromodel can capture the original system response well in a wide frequency range up to 10 GHz.

Generally, if a minimum function is expressed as the ratio of two nth-degree polynomials without common factor, it decreases two degrees during every synthesis step with an increased one stage order of a RLCM ladder. Usually, by increasing the ladder stage, we increase the model order and capture more poles. For example, this procedure captures two poles whenever we synthesize one more stage of the ladder.

Brune's synthesis needs realization of an equivalent coupled inductor to enforce passivity. Bott–Duffin's method [13], on the other hand, avoids the use of mutual coupled inductors. But a stiff price is paid in the extra number of RLC components and, hence, is not suitable for simple RLCM macromodel realization.

Note that for general transfer function realization, two-port network realization is required, as transfer functions are not positive real in general and one-port network realization cannot be applied directly [127]. In the following section, we present a general one-port and multi-port network synthesis method for admittance network realization.

11.2 General multi-port network realization

In this section, we introduce a new one-port and multi-port network synthesis approach. The new approach is motivated by the fact that the realized RLC circuits do not need to be physically implemented, as our goal is behavioral modeling. As a result, we allow the non-physical (negative) values of RLC elements. Such a *relaxation* can significantly simplify the realization process and render the process free of errors. On the other hand, passivity is enforced at the admittance and impedance matrix level (the mathematical level). Hence, the realized circuits are still passive, which was validated in [94].

11.2.1 General relaxed one-port network realization

We start with one-port network realization. The approach is based on one-port network realization using Foster's form, which can directly synthesize the one-port admittance function [127].

To synthesize the one-port circuit from the driving-point admittance rational function $Y(s)$, we first rewrite it in Foster's canonical form [127]:

$$Y(s) = sY_\infty + Y_0 + \sum_{m=1}^{M} \frac{a_m}{s - p_m} + \sum_{n=1}^{N} \left(\frac{a_n}{s - p_n} + \frac{a_n^*}{s - p_n^*}\right), \tag{11.17}$$

where we expand the rational function into the partial fraction form with N conjugate poles p_n and M real poles p_m.

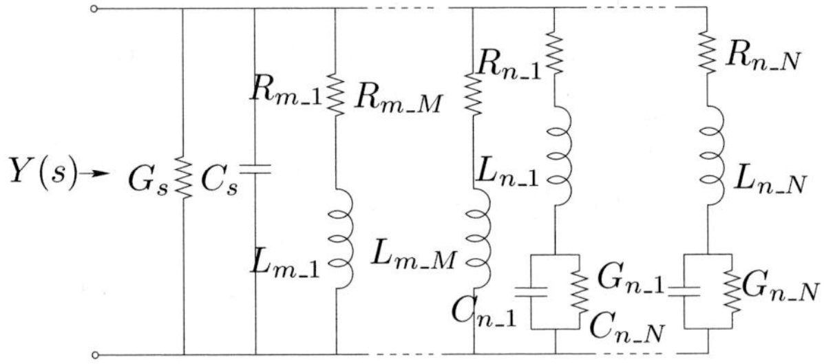

Figure 11.8 One-port Foster admittance realization. Reprinted with permission from [94] (c) 2006 IEEE.

The admittance function in the Foster's canonical form can then be synthesized by an equivalent circuit in Figure 11.8 with the following relations to determine R, L, C, G elements:

$$\begin{aligned}
G_s &= Y_0, \qquad C_s = Y_\infty; \\
R_{m_m} &= \frac{1}{a_m}, \qquad L_{m_m} = -\frac{p_m}{a_m}; \\
L_{n_n} &= \frac{1}{2\mathrm{Re}\{a_n\}}, \qquad L_{n_n}C_{n_n}|p_n|^2 = R_{n_n}G_{n_n} + 1, \\
\frac{G_{n_n}}{C_{n_n}} &= -\frac{\mathrm{Re}\{a_n p_n^*\}}{\mathrm{Re}\{a_n\}}, \qquad \frac{R_{n_n}}{L_{n_n}} = \frac{\mathrm{Re}\{a_n p_n^*\}}{\mathrm{Re}\{a_n\}} - 2\mathrm{Re}\{p_n\}.
\end{aligned} \tag{11.18}$$

Some existing works, like PRIME [82], require every complex pole pair to be physically realizable (every RCL element is positive), which is over-constrained and may lead to significant errors when unrealizable pole pairs are discarded or their residues are changed. In this approach, we *relax* those constraints by allowing some negative RLC elements.

Note that, in general, the synthesis may produce negative-valued circuit elements. These negative circuit elements will not affect the stability and passivity of the realized circuit when the original circuit matrix or rational function is passive [35, 39, 127] as the passivity is guaranteed at the mathematical level of the models and the realization is error-free and reversible.

11.2.2 Multiple-port network realization

For the passive multi-port order-reduced admittance matrix, we propose a general complete-graph structure (in case of full admittance matrix) to realize the admittance matrix based on one-port realization. Note that such a structure does not apply to the impedance matrices. Hence, our approach can only be applied to multi-port admittance matrix realization. But this does not indicate the restrictions as most RLC circuits can be written into admittance forms by using modified nodal analysis approach. In the following, we first illustrate how a 2-port network is realized and then we extend this concept for the general n-port network realization.

Given a 2×2 passive admittance matrix, which is obtained by the hierarchical model order reduction method,

$$Y_{2\times 2}(s) = \begin{bmatrix} y_{11}(s) & y_{12}(s) \\ y_{21}(s) & y_{22}(s) \end{bmatrix}, \tag{11.19}$$

it can be realized exactly, using the Π-structure template shown in Figure 11.9, where each branch admittance will be realized by the one-port Foster's expansion method shown in Figure 11.8. Based on this template, such a realization can be easily extended to the multi-port case.

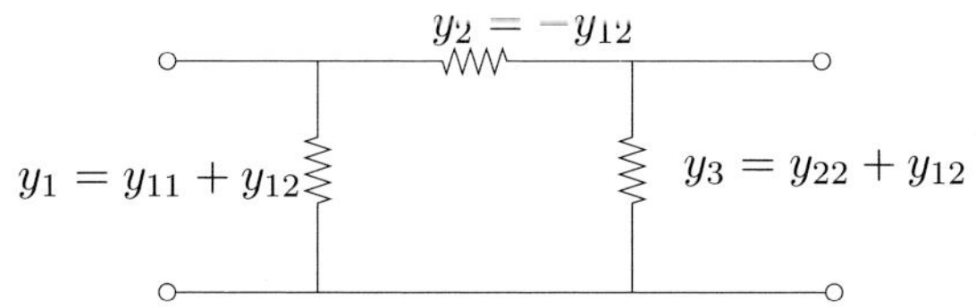

Figure 11.9 General two-port realization Π model. Reprinted with permission from [94] (c) 2006 IEEE.

$$Y_{N\times N}(s) = \begin{bmatrix} y_{11}(s) & \cdots & y_{1n}(s) \\ \vdots & \ddots & \vdots \\ y_{N1}(s) & \cdots & y_{NN}(s) \end{bmatrix}. \tag{11.20}$$

Generally, for a reduced N-port network with a full $N \times N$ admittance matrix as shown in (11.20), the realized network will be a complete graph where each branch represents an admittance that is realized by the one-port realization method. For instance, Figure 11.10 shows a realization of a synthesized six-port network. The branch admittance of the mth port branch (the branch between the port and ground) is the sum of all the mth row admittances, and the admittance of the branch between the port and any other port is the negative value of the corresponding admittance.

Notice that the new realized circuits automatically preserve the reciprocity of the original circuit matrix as it requires the admittance to be symmetric.

11.3 Multi-port non-reciprocal circuit realization

In this section, we discuss methods to realize non-symmetric circuit matrices, which are often the case for active circuits due to the presence of controlled sources.

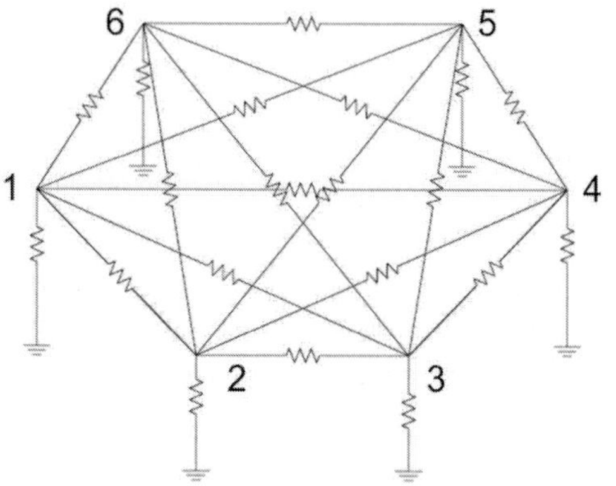

Figure 11.10 Six-port realization based on Π-structure. Reprinted with permission from [94] (c) 2006 IEEE.

In this section, we show how to realize a general non-symmetric (non-reciprocal) $n \times n$ admittance matrix into a macro-model in the form of RLC and controlled elements, which can be accepted by general SPICE-like simulators in both frequency and time-domain simulation.

Our approach is still based on the relaxed one-port realization approach. For one-port network, the realization process is the same as shown in Section 11.2. The difference lies in the multiple-port realization, as shown below.

To realize a general $n \times n$ non-reciprocal admittance matrix, we propose a general complete-graph structure (for the full admittance matrix) to realize the admittance matrix based on the one-port realization. In the following, we first illustrate how a 2-port network is realized and then we extend this algorithm for the general n-port network realization.

Given a 2×2 non-symmetric admittance matrix as shown in (11.21),

$$Y_{2\times 2}(s) = \begin{bmatrix} y_{11}(s) & y_{12}(s) \\ y_{21}(s) & y_{22}(s) \end{bmatrix}, \tag{11.21}$$

where $y_{12}(s) \neq y_{21}(s)$, the admittance matrix can be realized exactly by using the circuit template shown in Figure 11.11, where each branch admittance will be realized by the one-port Foster's expansion method shown in Figure 11.8.

Notice that the non-symmetric admittance is realized using two voltage-controlled voltage sources (VCVS), EV_1 and EV_2, and two current controlled current sources (CCCS), FI_{12} and FI_{21}. Both VCVS and CCCS have unit transfer gain. For instance, to realize $y_{12}(s)$, which represents the transconductor that leads

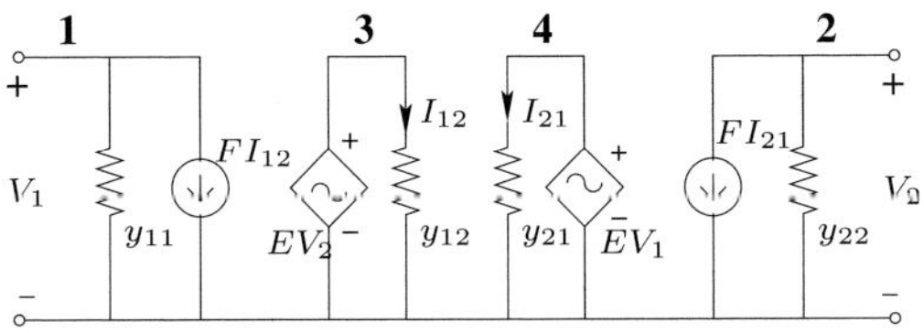

Figure 11.11 General two-port non-reciprocal active realization. Reprinted with permission from [73] (c) 2005 IEEE.

to the current injected into node 1 due to voltage at node 2, the VCVS EV_2 first transforms the voltage at node 2 (V_2) into the node 3 ($V_2 = V_3$). Then V_3 drives the admittance y_{12} to generate the current I_{12}, which then drives a CCCS FI_{12} to inject the same amount of current into node 1. Realization of $y_{21}(s)$ can be explained in a similar way.

For a general $n \times n$ non-symmetric admittance matrix, we can realize each pair of ports using the aforementioned two-port realization method until all the pairs of ports are realized. The resulting circuit will have a complete graph structure (for the full admittance matrix). But the non-symmetric property will be preserved during the realization process.

11.4 Numerical examples

In this section, we present two examples. The first example is a one-port passive network and its network function $Y_{1-port}(s)$ in the partial fraction form with four system poles is given below:

$$\begin{aligned}Y_{1-\text{port}}(s) = &\frac{3.4160 \times 10^7 + 1.1311 \times 10^6 i}{(s + 5.0661 \times 10^8 - 1.3497 \times 10^{10} i)} \\ &+ \frac{3.4160 \times 10^7 - 1.1311 \times 10^6 i}{(s + 5.0661 \times 10^8 + 1.3497 \times 10^{10} i)} \\ &+ \frac{2.1699 \times 10^8 + 6.9784 \times 10^6 i}{(s + 1.9829 \times 10^9 - 7.905 \times 10^{10} i)} \\ &+ \frac{2.1699 \times 10^8 - 6.9784 \times 10^6 i}{(s + 1.9829 \times 10^9 + 7.905 \times 10^{10} i)}.\end{aligned} \tag{11.22}$$

The frequency-domain responses of this circuit and the realized circuit are shown in (11.12). The two responses are the same.

Secondly, we present the realization results for a two-port passive network $Y_{2-\text{port}}(s)$ in partial fraction form with six system poles, as shown below:

$$Y_{2-\text{port}}(s) = \begin{bmatrix} Y_{11}(s) & Y_{12}(s) \\ Y_{12}(s) & Y_{22}(s) \end{bmatrix}, \tag{11.23}$$

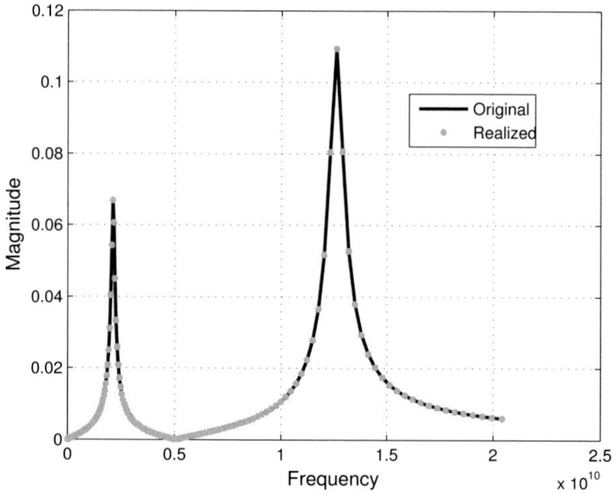

Figure 11.12 Comparison between the transfer function $Y_{1-port}(s)$ and its circuit realization.

where

$$
\begin{aligned}
Y_{11}(s) = &\frac{-1.5280 \times 10^7 - 1.4916 \times 10^8 i}{(s + 5.9850 \times 10^{10} - 2.2440 \times 10^{10} i)} + \\
&\frac{-1.5280 \times 10^7 + 1.4916 \times 10^8 i}{(s + 5.9850 \times 10^{10} + 2.2440 \times 10^{10} i)} + \\
&\frac{2.3594 \times 10^7 + 1.7770 \times 10^7 i}{(s + 7.8192 \times 10^9 - 3.9525 \times 10^{10} i)} + \\
&\frac{2.3594 \times 10^7 - 1.7770 \times 10^7 i}{(s + 7.8192 \times 10^9 + 3.9525 \times 10^{10} i)} + \\
&\frac{-1.2283 \times 10^7 - 2.3393 \times 10^7 i}{(s + 9.8407 \times 10^9 - 8.2723 \times 10^9 i)} + \\
&\frac{-1.2283 \times 10^7 + 2.3393 \times 10^7 i}{(s + 9.8407 \times 10^9 + 8.2723 \times 10^9 i)},
\end{aligned}
$$

$$
\begin{aligned}
Y_{12}(s) = {} & \frac{-7.3490\times10^{7}-1.3550\times10^{8}i}{(s+5.9850\times10^{10}-2.2440\times10^{10}i)} + \\
& \frac{-7.3490\times10^{7}+1.3550\times10^{8}i}{(s+5.9850\times10^{10}+2.2440\times10^{10}i)} + \\
& \frac{3.8682\times10^{7}+1.1596\times10^{7}i}{(s+7.8192\times10^{9}-3.9525\times10^{10}i)} + \\
& \frac{3.8682\times10^{7}-1.1596\times10^{7}i}{(s+7.8192\times10^{9}+3.9525\times10^{10}i)} + \\
& \frac{-5.9312\times10^{6}+2.2050\times10^{7}i}{(s+9.8407\times10^{9}-8.2723\times10^{9}i)} + \\
& \frac{-5.9312\times10^{6}-2.2050\times10^{7}i}{(s+9.8407\times10^{9}+8.2723\times10^{9}i)},
\end{aligned}
\tag{11.24}
$$

$$
\begin{aligned}
Y_{21}(s) = {} & \frac{-7.3490\times10^{7}-1.3550\times10^{8}i}{(s+5.9850\times10^{10}-2.2440\times10^{10}i)} + \\
& \frac{-7.3490\times10^{7}+1.3550\times10^{8}i}{(s+5.9850\times10^{10}+2.2440\times10^{10}i)} + \\
& \frac{3.8682\times10^{7}+1.1596\times10^{7}i}{(s+7.8192\times10^{9}-3.9525\times10^{10}i)} + \\
& \frac{3.8682\times10^{7}-1.1596\times10^{7}i}{(s+7.8192\times10^{9}+3.9525\times10^{10}i)} + \\
& \frac{-5.9312\times10^{6}+2.2050\times10^{7}i}{(s+9.8407\times10^{9}-8.2723\times10^{9}i)} + \\
& \frac{-5.9312\times10^{6}-2.2050\times10^{7}i}{(s+9.8407\times10^{9}+8.2723\times10^{9}i)},
\end{aligned}
$$

$$
\begin{aligned}
Y_{22}(s) = {} & \frac{-6.5196\times10^{8}-5.9276\times10^{8}i}{(s+5.9850\times10^{10}-2.2440\times10^{10}i)} + \\
& \frac{-6.5196\times10^{8}+5.9276\times10^{8}i}{(s+5.9850\times10^{10}+2.2440\times10^{10}i)} + \\
& \frac{5.5074\times10^{7}-8.9274\times10^{6}i}{(s+7.8192\times10^{9}-3.9525\times10^{10}i)} + \\
& \frac{5.5074\times10^{7}-8.9274\times10^{6}i}{(s+7.8192\times10^{9}+3.9525\times10^{10}i)} + \\
& \frac{8.2948\times10^{6}-1.5720\times10^{7}i}{(s+9.8407\times10^{9}-8.2723\times10^{9}i)} + \\
& \frac{8.2948\times10^{6}+1.5720\times10^{7}i}{(s+9.8407\times10^{9}+8.2723\times10^{9}i)}.
\end{aligned}
$$

The frequency responses for $Y_{12}(s)$ and $Y_{22}(s)$ are shown in the Figure 11.13 and Figure 11.14. Again, the realized circuit has the exact responses of the original circuit.

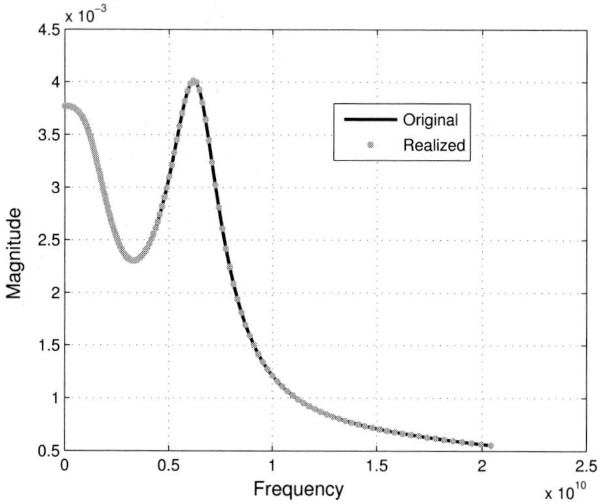

Figure 11.13 Comparison between the transfer function $Y_{12}(s)$ and its circuit realization.

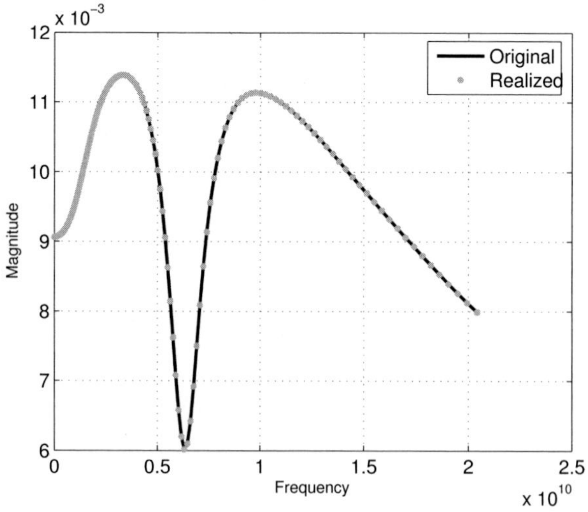

Figure 11.14 Comparison between the transfer function $Y_{22}(s)$ and its circuit realization.

11.5 Summary

In this chapter, we have presented two network realization approaches for generating SPICE-compatible circuit netlist from the reduced admittance or impedance matrices. The first method, Brune's method, is based on the classic network synthesis technique, which guarantees that the realized circuits are physically implemented. But we showed that such physical implementation is not necessary from the circuit simulation's perspective. Based on this observation, we presented a more simple yet efficient multi-port network realization approach, which is based on the relaxed one-port Foster network synthesis method. Our new approach allows for non-physical RLC elements, the passivity, however, is ensured at the admittance or impedance matrix level, thus, the realized circuits are still passive. We also extended this method to realize non-symmetric circuit matrices, which typically come from active linear or linearized circuits.

12 Reduction for multi-terminal interconnect circuits

In this chapter, we study the model order reductions on interconnect circuits with many terminals or ports. We show that projection-based model order reduction techniques are not very efficient for those circuits. We then present an efficient reduction method which combines projection-based MOR with a frequency domain fitting method to produce reduced models for interconnect circuits with large terminals.

12.1 Introduction

Krylov subspace projection methods have been widely used for model order reduction, owing to their efficiency and simplicity for implementation [32,37,85,91,113]. Chapter 2 has a detailed review of those methods.

One problem with the existing projection-based model order reduction techniques is that they are not efficient at reducing circuits with many ports. This is reflected in several aspects of the existing Krylov subspace algorithms like PRIMA [85]. First, the time complexity of PRIMA is proportional to the number of ports of the circuits as moments excited by every port need to be computed and matrix-valued transfer functions are generated. Second, the poles of the reduced models increase linearly with the number of ports, and this makes the reduced models much larger than necessary. The fundamental reason is that all the Krylov-based projection methods are working directly on the moments, which contain the information of both poles and residues for the corresponding transfer function. To deal with more ports, we have more transfer functions and thus more poles and residues to compute. However, poles among different transfer functions are the same for the same circuits as poles are characteristics of a system. But projection-based methods cannot take advantage of this as they operate directly on moments. As more residues are computed for more transfer functions, more poles also generated. However, generating more poles does not always help to improve the accuracy of the reduced models, as more block moments are not always matched as the number of poles increases. As a result, projection-based methods lead to larger reduced models than necessary when the number of ports is larger.

One way to resolve this problem is by means of port reductions. Recent work by

Feldmann *et al.* [31] exploited the port dependence to reduce the number of ports under some error metric constraints. This work, however, is limited to circuits with main similar ports and it cannot be applied to general linear circuits with many independent ports. Another approach, which is also amenable for circuits with multiple ports, is the hierarchical model order reduction method [93,121]. But hierarchical model order reduction is also numerically unstable, except for reducing tree circuits [121]. The improved version can produce more accurate models at the expense of multi-frequency point expansions [93].

In this chapter, we present an efficient model order reduction method that overcomes the difficulty associated with subspace projection-based MOR methods for reducing circuits with many ports. The basic idea of the new method is to compute separately the poles and residues of the transfer functions in the reduced admittance matrices. This can be achieved, first, by applying traditional subspace projection methods to compute the poles and then using hierarchical symbolic analysis for computing frequency responses of admittances to determine the residues of transfer functions. Since traditional projection-based MOR is used only for computing the poles, we only need to compute the poles necessary for the accuracy requirements. Finally to ensure the passivity of the reduced model, a convex programming based optimization is applied. The numerical results show that the new method can lead to much smaller reduced-model sizes for a given frequency range or much higher accuracy given the same model sizes than subspace projection based methods although at a higher computation cost.

12.2 Problems of subspace projection-based MOR methods

In this section, we briefly review the Krylov subspace based projection methods and point out their weakness for reducing circuits with multiple ports. Detailed discussions of Krylov subspace methods can be found in Chapter 2.

We look at the most representative Krylov subspace method, PRIMA [85]. Without loss of generality, a linear m-port electrical circuit can be expressed as

$$\begin{aligned} C\dot{x}_n &= -Gx_n + Bu_m \\ i_m &= L^T x_n, \end{aligned} \tag{12.1}$$

where x is the vector of state variables and n is the number of state variables, m is the number of independent sources specified as ports, C, G are matrices for conductance and storage elements, and B and L indicate input and output ports; typically $B = L$ for interconnect circuits and both of them are $n \times N$ matrices; N is the number of terminals (as input and output ports).

Define $A = -G^{-1}C$, $A \in \Re^{n\times n}$ and $R = G^{-1}B$, $R = [r_0 r_1 ... r_m] \in \Re^{n\times m}$. The y-parameter matrix after Laplace transformation is $Y(s) = L^T(G + sC)^{-1}B = L^T(I_n - sA)^{-1}R$ where I_n is the $n \times n$ identity matrix. The block moments of

$Y(s)$ are defined as the coefficients of the Taylor expansion of $Y(s)$ around $s = 0$:

$$Y(s) = M_0 + M_1 s + M_2 s^2 + ..., \tag{12.2}$$

where $M_i \in \Re^{N\times N}$ and can be computed as $M_i = L^T A^i R$.

The idea of model order reduction is to find a compact system much smaller than the original system. The Krylov subspace-based method accomplishes this by projecting the original system onto a special subspace, which expands the same space as the block moments of the original system. Specifically, the Krylov subspace is defined as

$$\begin{aligned} Kr(A, R, q) = \text{colsp}[&R, AR, A^2R, ..., A^{k-1}R, \\ &A^k r_0, A^k r_1, ..., A^k r_l] \quad (12.3)\\ k = \lfloor q/m \rfloor, &\qquad l = q - km. \quad (12.4) \end{aligned}$$

For simplicity of expression, we assume that $q = m \times k$ in the following. In reality any k can be chosen. Then, PRIMA tries to find orthogonal matrix $X \in \Re^{n\times q}$ such that $\text{colsp}(X) = Kr(A, R, q)$. With

$$\tilde{C} = X^T C X \qquad \tilde{G} = X^T G X$$
$$\tilde{B} = X^T B \qquad \tilde{L} = X^T L,$$

the reduced system of size q is found as

$$\tilde{C}\dot{\tilde{x}}_n = -\tilde{G}\tilde{x}_n + \tilde{B}u_m, \tag{12.5}$$
$$i_m = \tilde{L}^T x_n. \tag{12.6}$$

The reduced systems have q poles, which are the dominant poles of the original systems. Notice that the order of block moments k is related to the q by $k = \lfloor q/m \rfloor$. To match the first block moment ($k = 1$), we need at least m moments. For every one order of block moment increase, we need to add m additional poles. This is a highly inefficient reduction process.

We use the following practical industry interconnect circuits to further illustrate this problem. The circuit has 3 ports and 219 nodes. Figure 12.1 shows the frequency response of one of the nine transfer functions. With three block moments, PRIMA matches the exact response up to 3 GHz fairly well. However, to generate the Krylov subspace in (12.4), we need nine columns, which indicates that the reduced system size is nice, i.e., it has nine poles. But with the method to be presented in this chapter, we can actually obtain a reduced system of the same size that matches the original system response up to 100 GHz and still preserves passivity as PRIMA does. As the number of ports grows, generating a certain number of block moments becomes more and more expensive and the reduced system becomes much larger. However, in real industrial interconnect circuits, subcircuits can easily have ten to a hundred ports, and substrates can even be modeled with hundreds of ports. For this type of circuit, the Krylov subspace projection method will be extremely inefficient.

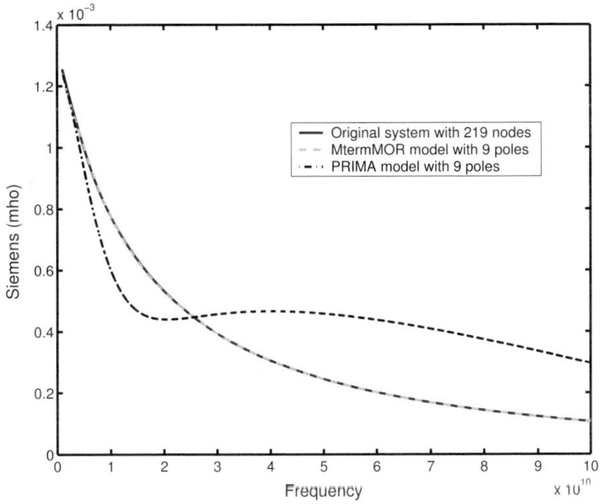

Figure 12.1 Frequency response of the three-input circuit. Reprinted with permission from [76] (c) 2006 IEEE.

12.2.1 *Transfer functions and dominant poles*

In this section, we show that all the admittance rational elements in the reduced models of a multi-input and multi-output (MIMO) system share the same set of dominant poles. This is also true for other types of circuit formulation (like impedance etc).

Let us define matrix $Y(s) = G + sC$ and assume $L = B$. The transfer function (admittance form) of the state space equation in (12.1) is

$$Y_{N\times N}(s) = B^T(G + sC)^{-1}B \tag{12.7}$$

$$= \frac{B^T[\Delta_{u,v}]^T R}{\det(Y(s))} \tag{12.8}$$

$$= \begin{bmatrix} y_{11}(s) & y_{12}(s) & \dots & y_{1N}(s) \\ y_{21}(s) & y_{22}(s) & \dots & y_{2N}(s) \\ \vdots & \vdots & \ddots & \vdots \\ y_{N1}(s) & y_{N2} & \cdots & y_{NN}(s) \end{bmatrix}, \tag{12.9}$$

where $y_{ij}(s)$ is the admittance rational transfer function between ports i and j and can be written as

$$y_{ij} = \frac{b_{ij,1} + b_{ij,2}s + \dots + b_{ij,m}s^m}{a_1 + a_2 s + a_3 s^2 + \dots + a_n s^n}, \tag{12.10}$$

where

$$a_1 + a_2 s + a_3 s^2 + \dots + a_n s^n = \det(Y(s)), \tag{12.11}$$

and $[\Delta_{u,v}]^T$ is the *adjoint* matrix of $Y(s)$. The dominator polynomial is common to all the y_{ij}. As a result, all the admittance elements y_{ij} share the same set of poles, which are the reciprocals of the eigenvalues of $Y(s)$.

During the reduction process, we can only obtain the approximate admittance elements $\bar{y}_{ij}$, which match the original admittance y_{ij} in terms of moments up to the order of $2q$ via explicit or implicit moment-matching methods, as mentioned in Chapter 2. Therefore, according to (2.34), if we have $2q$ moments, which are identical to the moments of original y_{ij}, then the computed denominator polynomial with order of q (via (2.34)),

$$\bar{a}_1 + \bar{a}_2 s + \bar{a}_3 s^2 + ... + \bar{a}_q s^{q+1} \tag{12.12}$$

will be the same as the truncated denominator polynomial of (12.11), i.e.,

$$a_i = \bar{a}_i, \quad \text{for } i = 1, ..., q \tag{12.13}$$

This is true for all the denominators of $\bar{y}_{ij}$, which should be the same. As a result, $\bar{y}_{ij}$ share the same set of the dominant poles after the reduction by the moment matching processes.

This observation motivates us to compute the poles and residues of all y_{ij} separately to reduce the model sizes and improve the reduction process efficiency. This technique is especially useful when we have large terminal count, thus large number of y_{ij}. Also, we apply Krylov subspace methods to compute the poles, which are numerically stable for computing sufficient numbers of poles. Next, we present the new reduction algorithm.

12.3 Model order reduction for multiple-terminal circuits: MTermMOR

In this section, we present an efficient model-reduction algorithm, which is suitable for circuits with multiple ports.

The key idea of the new method is to compute the poles and residues of the rational admittances (transfer functions) separately. In this way, we can avoid the generation of unnecessary poles as in the traditional subspace projection methods.

The new method consists of three steps to produce compact passive models for passive interconnect circuits.

Algorithm: MTermMOR (multi-terminal model order reduction)

1. Compute the dominant poles using a subspace projection-based method.
2. Compute the frequency responses of all the rational admittances for the desired frequency ranges by symbolic hierarchical analysis.
3. Apply the convex programming based method to find the residues of each rational admittance and ensure the passivity of the reduced admittance matrix.

Since the poles and residues are computed separately, we can only generate dominant poles; this impacts on the interested frequency range. As a result, only a few poles are typically required. Next, we will give each reduction step in detail.

12.3.1 *Computation of system poles by subspace-projection methods*

In this section, we show how to efficiently and reliably compute the dominant poles in rational admittance elements in (12.9). Instead of using an explicit moment matching method like AWE [91], which suffers numerical issues for computing high order moments, we use Krylov subspace methods, reviewed in Section 12.2, which perform the implicit moment matching, and are numerically stable for computing high order moments, and thus a sufficient number of poles for the reduction process.

Since we are only interested in the poles of the circuits, we do not need to perform full-blown model order reduction. Instead, we can only perform MOR on SISO systems or on MIMO systems with very low orders. As a result, the computation costs are no longer associated with the number of ports. This is because all the rational admittance elements among different ports share the same set of poles as shown in Subsection 12.2.1 and we only need to compute only the rational admittances $\bar{y}_{ij}$ in (12.9).

Therefore, we have two ways to compute the dominant poles in the subspace projection framework. The simplest way is to compute the moments from just one input (any one port) until the sufficient number of moments are obtained. In other words, we transform the pole computation problem into model order reduction problem on an SISO system. The second method obtains the moments from different inputs. If we use the moments from all m ports, we obtain the block moments as used in PRIMA. But we may only need a few orders of block moments as the poles generated will be km, where k is the order of block moments for the accuracy requirement. This idea is similar to the multiple-moment-matching method, where moments from different poles are used to compute the poles via the explicit moment matching method [54].

After the q moments are generated either from one input or multiple inputs, the projection method X is generated and the reduced system is obtained by the congruence transformation as shown in (12.5) and (12.6). The dominant q poles can be found by eigen-decomposition of the matrix $\tilde{G}^{-1}\tilde{C}$, whose eigenvalues are reciprocals of the dominant poles:

$$p_i = -\frac{1}{\lambda_i}, \tag{12.14}$$

where p_i and λ_i are the ith pole and eigenvalue.

Moreover, by applying the pole selection rule in [93], we actually prune out a great portion of poles that have little impact in the frequency range of interest. Specifically, in contrast to many analog circuits, where most poles are carefully designed to lie on the real axis, the poles of interconnect circuits are usually complex poles. As shown in [19], poles that dominate the transient response in interconnect circuits are those near the imaginary axis with large residues, and as observed in [93], the frequency range on which a pole has a major impact is closely related to its imaginary part: the pole has peak responses at frequency identical to the value of its imaginary part divided by 2π. With these observations we can eliminate poles that have little impact on the frequency range of our interest. Indeed as shown in

Figure 12.2, we can use only four of the nine poles in a example circuit and still match the exact response very well. The model system size is reduced from 219 to 4 but with very good accuracy as it matches with the original system up to 100 GHz. This example shows that there exists great potential to further reduce the model sizes from subspace projection-based methods while maintaining the same or better accuracy.

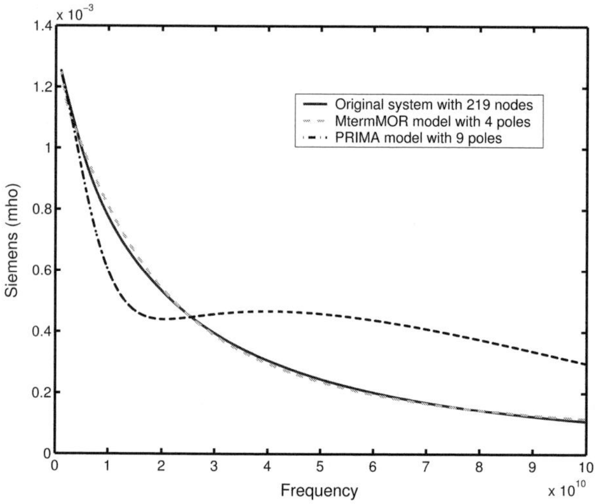

Figure 12.2 Frequency response of the three-input circuit with different approximation. Reprinted with permission from [76] (c) 2006 IEEE.

12.3.2 *Symbolic hierarchical analysis for admittance response*

Once the poles are computed, we need to compute the residues of each rational admittance function. For this propose, we need to know the frequency response of each rational admittance function in the reduced admittance matrix $Y_{k\times k}(s)$. This can be computed efficiently by using symbolic hierarchical analysis [126]. In the following, we briefly review the hierarchical symbolic analysis first. Then we compare it with SPICE for frequency domain admittance response computation.

For a given circuit formulated in the modified nodal analysis (MNA) as $MX = b$, it can also be rewritten in the following form (Schur's decomposition):

$$\begin{bmatrix} M^{II} & M^{IB} & 0 \\ M^{BI} & M^{BB} & M^{BR} \\ 0 & M^{RB} & M^{RR} \end{bmatrix} \begin{bmatrix} x^I \\ x^B \\ x^R \end{bmatrix} = \begin{bmatrix} b^I \\ b^B \\ b^R \end{bmatrix}. \tag{12.15}$$

The matrix, M^{II}, is the *internal* matrix associated with the internal variable vector x^I.

The aim of hierarchical reduction is to eliminate all the variables in x^I, and transform (12.15) into the following reduced set of equations:

$$\begin{bmatrix} M^{BB*} & M^{BR} \\ M^{RB} & M^{RR} \end{bmatrix} \begin{bmatrix} x^B \\ x^R \end{bmatrix} = \begin{bmatrix} b^{B*} \\ b^R \end{bmatrix}, \tag{12.16}$$

where $M^{BB*} = M^{BB} - M^{BI}(M^{II})^{-1}M^{IB}$ and $b^{B*} = b^B - M^{BI}(M^{II})^{-1}b^I$. Suppose that the number of internal variables is t, and the number of boundary variables is m. Then each matrix element in M^{BB*} and b^{B*} can be written in the following expanded forms:

$$a_{u,v}^{BB*} = a_{u,v}^{BB} - \frac{1}{\det(M^{II})} \sum_{k_1,k_2=1}^{m} a_{u,k_1}^{BI} \Delta_{k_2,k_1}^{II} a_{k_2,v}^{IB}, \tag{12.17}$$

where $u, v = 1, ..., m$ and

$$b_u^{B*} = b_u^B - \frac{1}{\det(M^{II})} \sum_{k_1,k_2=1}^{m} a_{u,k_1}^{BI} \Delta_{k_2,k_1}^{II} b_{k_2}^I, \tag{12.18}$$

where $u = 1, ..., m$ and $\Delta_{u,v}$ is the first-order *cofactor* of $\det(M)$ with respect to $a_{u,v}$.

The aim of symbolic hierarchical analysis is to represent all the determinants like $\det(M^{II})$ and first-order cofactors like $\Delta_{k_2,k_1}^{II} a_{k_2,v}^{IB}$ and $\Delta_{k_2,k_1}^{II} b_{k_2}^I$ symbolically using the determinant decision diagram (DDD) based method. It was shown in [126] that this is as efficient as the direct solution methods like LU-decomposition methods and can be scalable to large circuits.

For our problem, we need to compute all the rational admittances in the reduced admittance matrix. For a system with m ports, we have m^2 admittance functions to compute. It turns out that there are huge advantages in using the symbolic hierarchical method over LU-based numerical methods. The reason is that the cost of computing all the rational admittances is almost similar to the cost of computing just one rational admittance because of the symbolic sharing among all the rational admittances.

We use the clock net with 219 nodes to illustrate this. Figure 12.3 shows the computation costs of the symbolic hierarchical method versus SPICE on this circuit for different numbers of ports. Notice that each entry in the admittance matrix is essentially a transfer function. For example, when computing the frequency response of the $(1, 2)$ entry, we need to apply a voltage source at port 1 and observe the current at port 2. A lot of information can be reused in our hierarchical symbolic analysis engine, but SPICE cannot take advantage of this and we need to change the input and output information for each admittance matrix entry. As a result the simulation time of SPICE increases linearly with the number of ports, as shown in Figure 12.3. But for the hierarchical analysis the computational cost increase is much smaller and the difference will become more significant, as more ports are used.

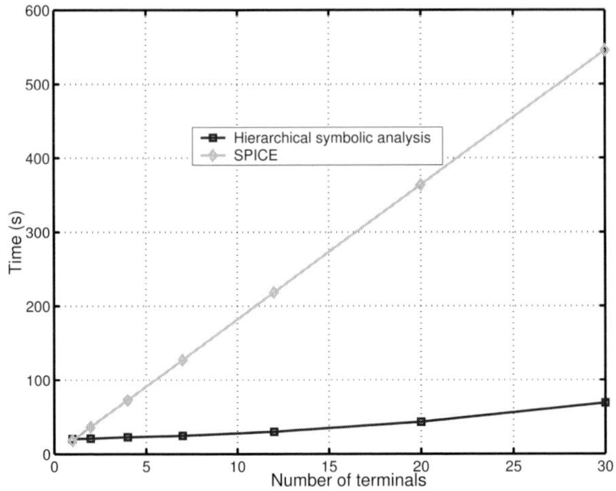

Figure 12.3 Comparison of computation cost for admittance. Reprinted with permission from [76] (c) 2006 IEEE.

12.4 Numerical examples

The new method has been implemented using C++ and MATLAB. We tested the algorithm on two real industry interconnect circuits from industry. The convex programming problem is solved using a standard optimization package, SeDuMi [118] in our passivity enforcement implementation. SeDuMi is powerful software package add-on for MATLAB, which allows one to solve optimization problems with linear, quadratic, and semi-definite constraints.

We use PRIMA as the Krylov subspace projection method for pole computation implemented in MATLAB. The frequency responses are computed using the symbolic hierarchical analysis tool [124, 126]. Then the convex programming tool, SeDuMi, takes in the poles and frequency response information and perform the optimization to produce the final reduced models in frequency-domain. The first interconnect circuit *clktree50*, is a clock-tree circuit in 180 nm technology. This network contains 210 nodes and 9 terminals.

In Figure 12.4, the PRIMA model's response is accurate at the frequencies up to 15 GHz. However the MTermMOR model's response matches the original circuit up to 50 GHz. Moreover, the MTermMOR model only contains five dominant poles compared with 90 poles in PRIMA model.

The similar results are obtained from another example *sram1026*, which is one bit line circuit from a SRAM circuit also in 160nm technology. This circuit includes 516 nodes and 11 terminals. The experimental results are shown in Figure 12.5. PRIMA's model using 44 poles is accurate up to 2 Ghz, while the MTermMOR model matches the original circuit's response up to 10 Ghz with just four poles. Both examples clearly demonstrate the advantage of the new method.

Although the new method is very effective for generating compact models, its

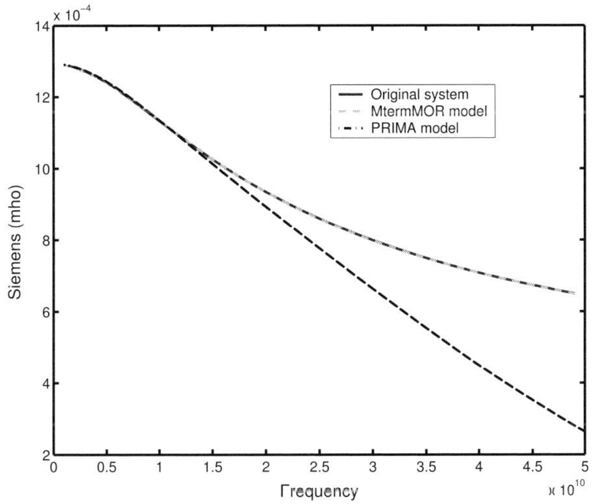

Figure 12.4 Frequency response comparison among the original circuit, PRIMA model, and MTermMOR model of the circuit *clktree50*. Reprinted with permission from [76] (c) 2006 IEEE.

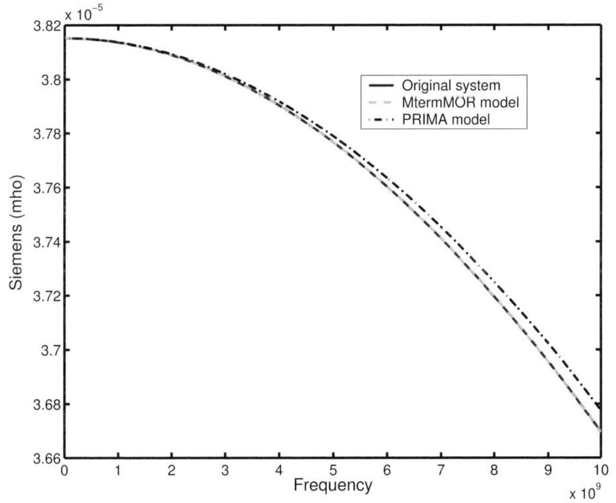

Figure 12.5 Frequency response comparison among the original circuit, PRIMA model, and MTermMOR model of the circuit *sram1026*. Reprinted with permission from [76] (c) 2006 IEEE.

capacity is limited by the convex programming step used. The current computing configurations (500 MB memory) allow optimization for circuits with about 15 ports in the MATLAB computing environment. With more memory, we expect that circuits with more ports can be handled. Also, the improved version of convex programming [21], which exploits the structures of the circuits, will also help to improve the efficiency of the convex programming algorithm, and we will look at those methods in the future.

12.5 Summary

In this chapter, we presented a new reduced modeling technique to reduce circuits with many ports. The new method overcomes the difficulty associated with subspace projection-based MOR methods for reducing circuits with many ports. Our contribution is new reduction flow, which computes the poles and residues of each transfer function separately in the reduced admittance matrices. The resulting reduction algorithm uses a much smaller number of poles to achieve the same accuracy than do subspace projection-based methods. To ensure the passivity, convex-programming-based optimization is used. The new method is a closed-loop method as it can produce models matching the desired frequency range precisely. The new method can lead to much smaller reduced models for a given frequency ranges or much higher accuracy given the same model sizes than subspace projection-based methods. Experimental results demonstrated the advantage of the new method over the subspace projection-based methods.

13 Passive modeling by signal waveform shaping

In this chapter, we exploit a new way of passive modeling interconnect circuits by so-called *signal waveform shaping*. Traditional passive modeling methods try to ensure that the reduced models are strictly passive using passive-preserving model order reduction methods or passivity enforcement optimization methods, as shown in Chapter 2 and Chapter 10. In this chapter, we show that passivity can also be achieved by slightly altering (shaping) the waveforms passing through the reduced models.

13.1 Introduction

Many model order reduction (MOR) techniques have been proposed in the past. The most efficient and successful one is based on subspace projection [32,37,85,91,113], which was pioneered by asymptotic waveform evaluation (AWE) algorithm [91] where explicit moment matching was used to compute dominant poles at low frequencies. After AWE, more numerical stable techniques were proposed [32, 37, 85, 113] by using implicit moment matching and congruence transform to produce passive models. A detailed review of Krylov subspace methods can be found in Chapter 2.

Another important development in linear MOR is the introduction of truncated balanced realization (TBR) methods, where the weak uncontrollable and unobservable state variables are truncated to achieve the reduced models [81, 87, 89, 131]. The TBR methods can produce nearly optimal models but they are more computationally expensive than projection-based methods. Also, TBR can produce passive models in its latest developments [87, 89, 131], which can be approximately viewed as generalized projection-based reduction methods [89]. The TBR methods are reviewed in Chapter 3.

Although projection-based MOR methods are very successful, their applications are mainly limited to RLC circuits. Many high-speed circuits, like RF surface acoustic wave (SAW) filters, spiral inductors, and high-speed transmission lines, are still modeled by using measured data (like scattering parameters) on account of many high frequency effects and the frequency dependency of circuit parameters. Another issue with projection-based MOR methods is that they become very inefficient for reducing circuits with many terminals in terms of both computational costs and reduced model sizes [60]. The main reason for the lost efficiency lies in the fact that with more terminals, more transfer functions must be computed and more poles will be used for each increased order of block moments, which is not necessary as poles are system information and should not depend on the terminals.

For generating general-purpose compact models from many measured and simulated data, fitting methods based on a least-square rational approximation in the frequency domain are still widely used [50]. One critical issue in such a modeling process is to preserve the passivity of the original system in the reduced models. Existing approaches, such as PRIME [82], enforce the passivity by physically realizing each pole and residue (or conjugate pole pair) term in the fractional form using Foster's synthesis method. If a pole/residue term cannot be realized, it is discarded. As a result, PRIME may lead to very large errors. The latest approach to this general passivity-enforcement problem is based on the convex programming (CP) approach using the state-space representation of the system [21]. The passivity is enforced using semi-definite constraints during a semi-definite (convex) optimization. But the CP-based method suffers very high computational costs and can optimize circuits with less than about 20 poles and 20 terminals in a typical computation setting (on the latest Intel Pentium-4 CPU with 1 GB memory).

In this chapter, we present a new passivity enforcement approach for general-purpose modeling of passive linear circuits. The new method is based on the observation that most interconnect circuits, such as clock trees, substrate, packing, RF passives, and transmission lines, are lossy and their frequency responses behave like a band-pass or low-pass filter in general. As a result, the models for those passive systems need not be passive for all frequencies, as required by traditional passivity-enforcement methods. Practically, they only need to be passive for a limited bandwidth, in which most of the signal energy is concentrated.

Instead of making the reduced models passive for all frequencies, the new method works on the signals going into the reduced models to enforce the passivity. The idea is to shape the waveforms of the signals slightly, such that the resulting spectra are band limited to the frequency range in which the reduced system is passive. As a result, the reduced models only need to be band-limited passive, which we call conditionally passive in this chapter, and can be achieved much more easily than traditional passivity for a reduced system. We present two approaches to band limit (shape) the waveforms. The first method is based on frequency-domain fast Fourier transform (FFT) and inverse FFT to shape the waveforms explicitly. The second method is based on insertion of passive low-pass filters (LPF) into the reduced models to shape the waveforms implicitly. For the second method, we analyze the delay and distortion effects introduced by using low-pass filters and present methods to mitigate the delay effects. Experimental results on several interconnect circuits demonstrate the effectiveness of the new methods.

13.2 Passivity and positive-realness

In this section, we review the standard passivity definition for a network function. Passivity is an important property of many physical systems. A passive network does not generate energy. If the reduced model loses its passivity, it may lead to unbounded responses in transient simulation, which means that new energy has been generated in this network.

Figure 13.1 shows a transient simulation result of a non-passive circuit under a sinusoidal excitation.

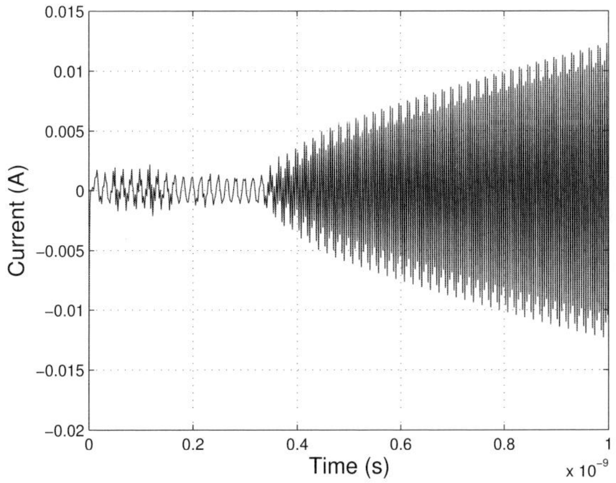

Figure 13.1 Transient response of a non-passive circuit.

Brune [12] has proved that the admittance and impedance matrix of an electrical circuit consisting of an interconnection of a finite number of positive R, positive C, positive L, and transformers are passive if and only if their rational functions are positive real. A network with admittance matrix function $\mathbf{Y}(s)$ is said to be positive real iff

$$\begin{aligned}&(1)\ \ \mathbf{Y}(s) \text{ is analytic, for } \mathrm{Re}(s) > 0\\&(2)\ \ \overline{\mathbf{Y}}(s) = \mathbf{Y}(\bar{s})\ , \text{ for } \mathrm{Re}(s) > 0\\&(3)\ \ \mathbf{Y}(s) + \mathbf{Y}(s)^H \geq 0\ , \text{ for } \mathrm{Re}(s) > 0.\end{aligned}$$

Condition (1) means that there are no unstable poles (poles lying on the right-half-plane (RHP) in the s-domain). Condition (2) refers to a system that has real response. And condition (3) is equivalent to the real part of $\mathbf{Y}(s)$ having a positive semi-definite matrix at all frequencies. In other words, the real parts of all the eigenvalues of $H(s)$ must be equal to or larger than zero. But condition (3) is difficult to satisfy, as it requires the checking of frequency responses from DC to infinity.

13.3 Conditional passivity and positive-realness

In this section, we analyze the relationship of a system's transient responses and its input signals in terms of passivity. We show that a non-passive system can still behave like a passive system when its input signals are band limited. Such systems can be defined as conditionally passive and their network functions are conditionally positive real.

In the previous section, we saw that for a passive system, its admittance matrix $Y(s)$ needs to be positive real ($\mathrm{Re}\{Y(s)\}$) is positive definite) for all frequencies. However, when $Y(s)$ is not positive real for some frequency ranges, will the system always exhibit non-passive behavior in the time domain, as shown in Figure 13.1? Actually the answer depends on the spectrum (energy) of the input signal. If the input signal is band-limited to the frequency range where the reduced model is positive real, then the system will still behave passively as the original system.

This can be illustrated by the following example. Figure 13.2 shows the frequency responses of a RLC circuit and its reduced model. The two circuits match well below 15 Ghz. Above 60 Ghz, the real part of the transfer function of the reduced model becomes negative as shown in Figure 13.2(a), which means that the system is no longer passive. When a sinusoidal input signal of 10 Ghz is applied to both systems, we get the exact responses in the time domain, as shown in Figure 13.3. However, if we apply a sinusoidal signal of 60 Ghz, the time-domain responses of the original system and the reduced system will be dramatically different, as shown in Figure 13.4. The response of the reduced system actually explodes.

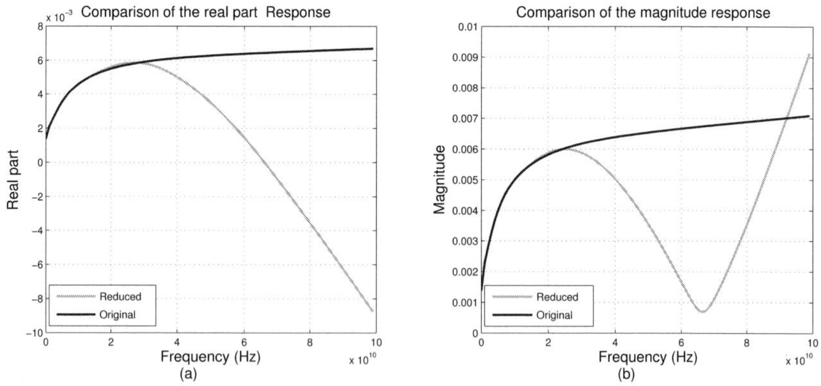

Figure 13.2 Frequency responses of a reduced model and its original RC circuit.

We note that many ideal input signals such as the Dirac delta function $\delta(t)$, unit step function $u(t) = 1, t \geq 0, u(t) = 0, t < 0$, and unit ramp function $f(t) = t, t \geq 0, f(t) = 0, t < 0$, have an infinite spectrum of frequencies. For example, for the Dirac delta function, $L(\delta(t)) = 1$, where $L(X)$ means taking the Laplace transform of function X. So $\delta(t)$ has a constant spectrum for all frequencies and it can easily make any non-passive system exhibit the non-passive behavior, as shown in [63].

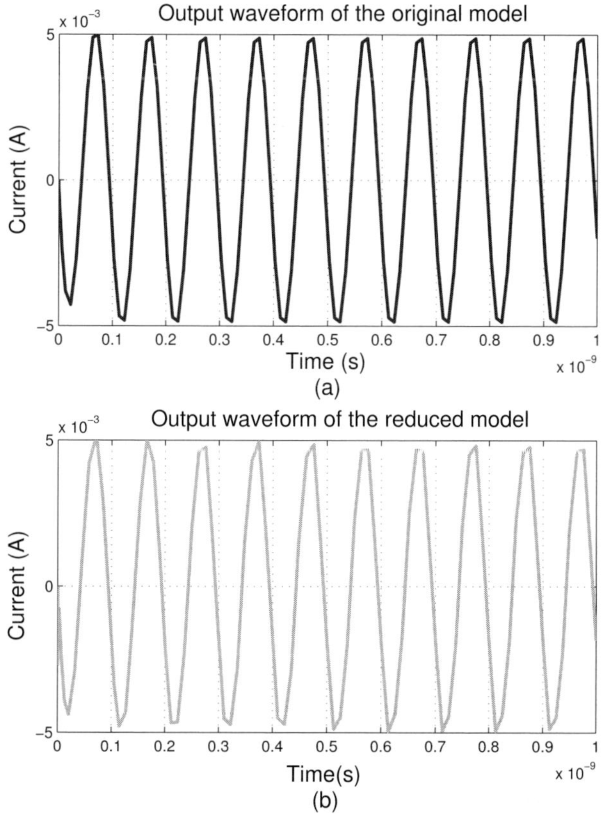

Figure 13.3 Transient responses of a reduced model and its original RC circuit for a 10 GHz input.

The Laplace transform of a unit step function and a unit ramp function are $1/s$ and $1/s^2$, respectively. When those signals are applied to a non-passive system, non-passive behavior can easily be observed as shown in [85].

However, such ideal signals do not exist in the real world. Most of the active transistors, passive interconnects, RF passive components, and transmission lines exhibit limited bandwidth due to unavoidable capacitive and other loss, which implies that signals generated by and propagated through those systems will bear limited bandwidth. This situation will become worse as we move to the sub-100 nm technologies. Hence for realistic signals, we can build a reduced system that is only passive for the given frequency range and the resulting system will still be passive as far as the simulation is concerned. For this purpose, we introduce conditional passivity and conditional positive-realness.

A network with admittance matrix function $\mathbf{Y}(s)$ is said to be conditionally

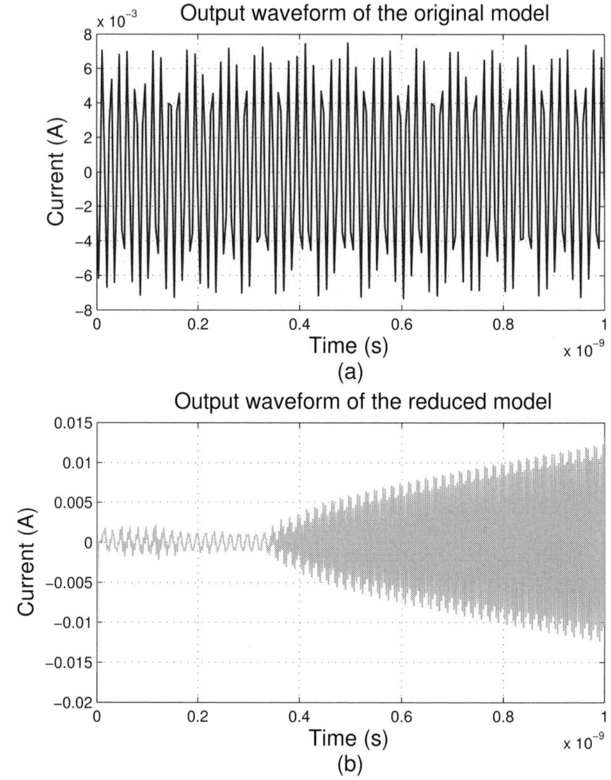

Figure 13.4 Transient responses of a reduced model and its original RC circuit for a 60 Ghz input.

positive real iff

$$\begin{array}{ll}(1)\ \ \mathbf{Y}(s) \text{ is analytic, for } \mathrm{Re}(s) > 0 & \\ (2)\ \ \overline{\mathbf{Y}}(s) = \mathbf{Y}(\bar{s})\ , \text{ for } \mathrm{Re}(s) > 0 & 0 \leq \mathrm{Im}(s) \leq 2\pi f_{\max} \\ (3)\ \ \mathbf{Y}(s) + \mathbf{Y}(s)^H \geq 0\ , \text{ for } \mathrm{Re}(s) > 0 & \end{array}$$

In other words, $\mathbf{Y}(s)$ will be positive real for the given frequency range $[0, f_{\max}]$.

The main benefit of a reduced system being conditionally passive is that conditional passivity can be much easier to achieve than strict passivity. Many existing frequency-domain rational fitting methods [50,78] can be used to do this with much more scalable computational costs than the convex programming method [21]. On the other hand, we put more constraints on the signals driving the conditionally passive systems: we need to make sure that the signal spectrum is band limited such that its bandwidth is within the positive real bandwidth of the reduced system. In the following section, we present two methods to achieve this requirement.

13.4 Passivity enforcement by waveform shaping

In this section, we discuss two methods to band limit a signal by slightly shaping its waveform. Note that based on the Fourier transform, if a signal is finite in the time domain, its spectrum extends to infinite frequency, and if its bandwidth is finite, its duration is infinite in the time domain. For a practical non-periodic time-limited signal such as a switching current in the signal line from transistor switching, one can never band limit such a signal from a strictly mathematical point of view. But practically we can make the out-of-band frequency energy sufficiently small compared with the in-band frequency energy such that the out-of-band energy will not stimulate the non-passive behavior of the system.

13.4.1 FFT and IFFT based waveform shaping

The first method is based on the fast Fourier transform (FFT) and inverse fast Fourier transform (IFFT). The idea is to first transform the original transient signal into the frequency domain. Since in the FFT (or the discrete Fourier transform, DFT), we treat the non-periodic signal as a periodic signal, the resulting signal's spectrum becomes discrete. Then we truncate those frequencies beyond $f_{\max}$, which is given by designers. After this, we perform the inverse FFT on the truncated spectrum to get the time-domain waveform of the shaped signal (we only take the waveform in one period). The whole process is illustrated in Figure 13.5 and the algorithm is outlined in Figure 13.6.

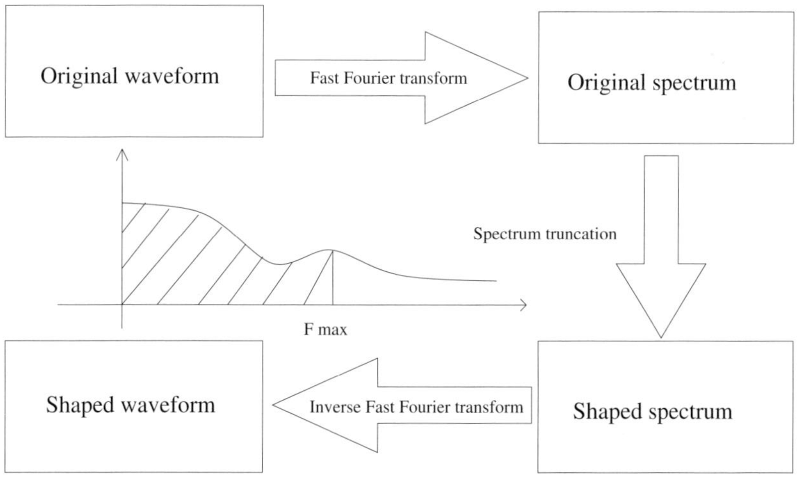

Figure 13.5 Algorithm flow of FFT-IFFT-based waveform shaping.

Figure 13.7(a) and (b) show a ramp signal and its spectrum. The shaped waveform with the cut-off frequency $f_{\max} = 10\,\text{Ghz}$ and the corresponding truncated spectrum are shown in Figure 13.7(c) and 13.7(d). The shaped waveform with the cut-off frequency $f_{\max} = 2\,\text{Ghz}$ and the corresponding truncated spectrum are

FFT_IFFT_WAVEFORMSHAPING
Sample input data with F_s;
Fast Fourier transform
$X_0(k) = \sum_{j=1}^{N} x_0(j) w_N^{(j-1)(k-1)}$;
Spectrum truncation
If $f_k < f_{\max}$ or $f_k > F_s - f_{\max}$,
$X_1(k) = X_0(k)$;
If $f_{\max} < f_k < F_s - f_{\max}$,
$X_1(k) = 0$;
Inverse fast Fourier transform
$x_1(j) = (1/N) \sum_{k=1}^{N} X_1(k) w_N^{-(j-1)(k-1)}$;
return vector: x_1 of length N;

Figure 13.6 Algorithm of FFT-IFFT-based waveform shaping.

shown in Figure 13.7(e) and (f). In general, the spectrum truncation does not significantly change the waveform characteristics such as delay and slew. As we truncate high-frequency components, the shaped waveform shows some undershoots and overshoots in Figure 13.7(e). Those small undershoots and overshoots do not affect the delay and timing of the shaped waveform, when it propagates through the reduced model. If we truncate the spectrum at a higher frequency such as 10 GHz, we find that the resulting waveform is almost the same as the original one, which is shown in Figure 13.7(c). This demonstrates that if the cut-off frequency is high enough, the distortion caused by truncating can be tolerated.

The drawback of the explicit waveform shaping method using FFT and IFFT is that it takes extra computational costs to process the signals. The computational cost is $O(n\log_2(n))$, where n is the number of sampling points.

13.4.2 Low-pass-filter based waveform shaping

The second method is based on implicit waveform shaping by adding passive low-pass filters between the input signal and the reduced system, as shown in Figure 13.8. In this way, we guarantee that the signals through the reduced system are band limited.

Notice that if we have a few input terminals (as for many interconnect circuits like clock trees or clock meshes), adding a few filters at those terminals will not increase the sizes of the reduced models significantly.

Since the filter can be passively realized by a LC ladder, it can be combined with the reduced model to function as a passive model. Therefore, we can conveniently use this new model in current simulation software such as SPICE.

However, we need to look at several issues associated with this method before we use it. First, the low-pass filter may also distort the input signals, as different frequency components may be delayed differently. Second, the introduction of a

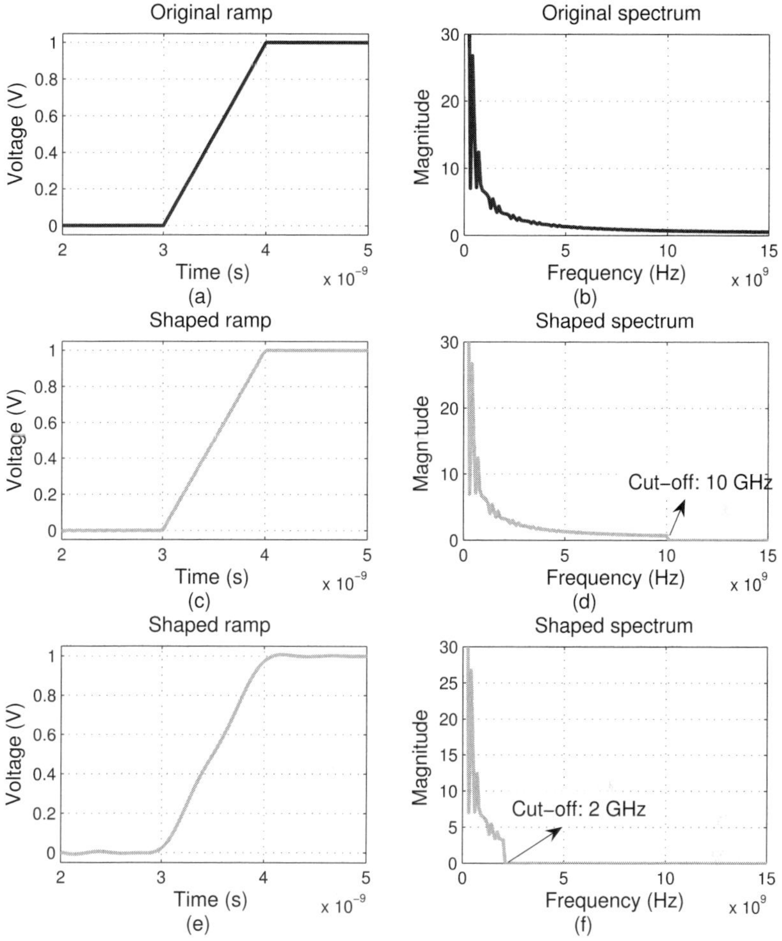

Figure 13.7 Ramp signal shaped at different frequencies.

low-pass filter will introduce delay. In the following, we discuss methods to mitigate those two problems.

Mitigation of distortion problems

The phase function and the resulting group delay function of a filter have profound time-domain ramifications as they have a direct effect on the waveform shape of the output signals. As a result, we choose the Bessel filter family for its good time-domain property. A Bessel filter has a linear phase characteristic over the passband of the filter, which implies a constant time delay over the pass-band of the filter (see Figure 13.9) so that the phase distortion in the filtering process can be avoided.

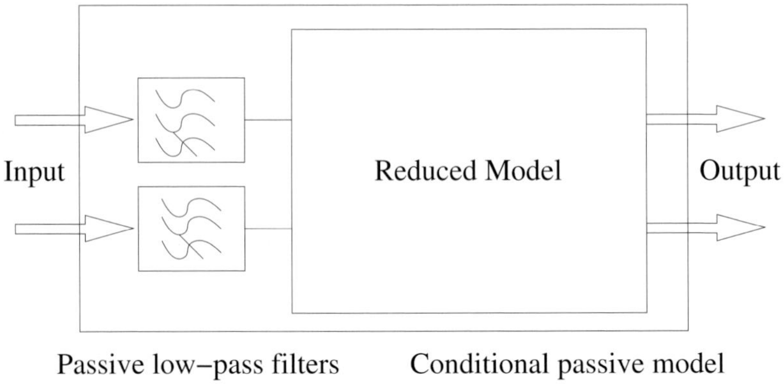

Figure 13.8 Low-pass-filter-based waveform shaping.

From Figure 13.9(a), we can see a constant time delay from DC to the normalized frequency one when the order (n) of the filter is higher than three. In addition, its step response exhibits negligible overshoot and ringing.

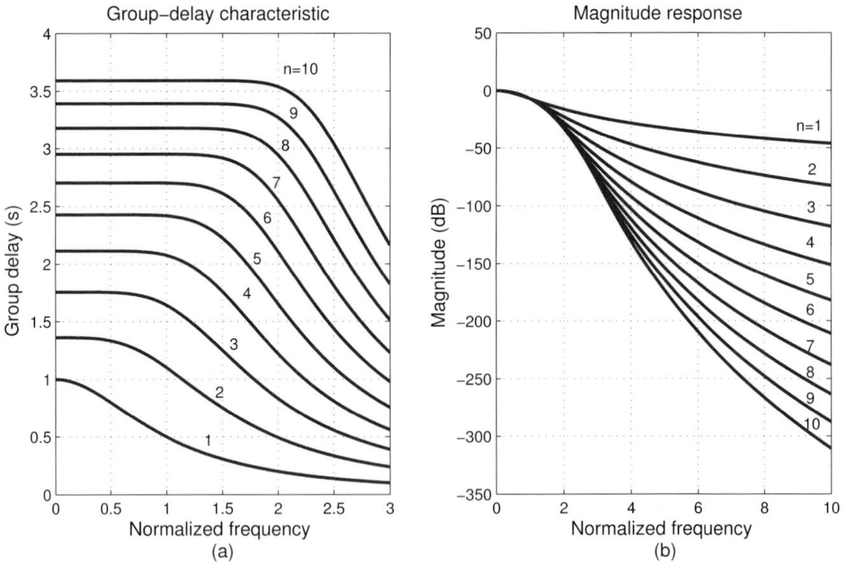

Figure 13.9 Group-delay characteristic and magnitude response for different order Bessel filters (normalized frequency).

However, a gradual roll-off (longer decay range) is the price we have to pay for a good time domain property. Fortunately, we can compromise it by increasing the order of the filter (see Figure 13.9(b)) at the cost of larger reduced models. Another way is to increase the passive frequency range so that the filter has sufficient reduction of spectrum (again at the cost of larger reduced models).

Mitigation of delay problems

Another issue we have to take into consideration is the time delay caused by the filter. Three factors can influence the time delay: the prototype, the order, and the cut-off frequency of a filter. Among them, the cut-off frequency is the dominant factor because the delay is inversely proportional to the cut-off frequency of a filter.

$$\text{Actual delay} = \frac{\text{Normalized delay}}{\text{Actual corner frequency (fc)}}. \tag{13.1}$$

For example, for the eighth order Bessel filter, the normalized delay is 2.703 s. If the cut-off frequency is as high as 20 GHz, the actual delay could be as small as 0.135 ns.

Hence, if the cut-off frequency is sufficiently high, the group delay caused by the filter can be made sufficiently small compared with the delay of the original circuit so that such a delay can be ignored.

13.5 Numerical examples

In this section, we present some experimental results on two interconnect circuits from industry designs (from Cadence Design Systems Inc.). Conditional passivity is achieved by using the minimum-square fitting method on the required transfer functions with poles computed from projection-based methods, such as PRIMA. This fitting method can make the reduced models accurate to the given maximum frequency and ensure the passivity of the models in the given frequency range.

The first example is a RC circuit with 210 nodes and 3 terminals. In this experiment, we used a steep square waveform as the input signal, as shown in Figure 13.10(a). We applied this signal to the original model, the reduced model, and the LPF (low-pass filter) based reduced model. The output waveforms of these three models are shown in Figure 13.10(b),(c),(d), respectively.

The reduced model is conditionally passive: the passivity of the model can only be preserved at the frequency range from DC to 15 GHz. Since a steep square waveform contains high-frequency components beyond this range, we can observe the erratic time-domain behavior caused by energy generated at high frequencies, as shown in Figure 13.10(c).

However, by eliminating those high-frequency components by LPF, the output waveform of the LPF-based reduced model (Figure 13.10(d)) matches the output waveform of the original model (Figure 13.10(b)) with little discrepancy. Therefore, the LPF-based reduced model can function as a passive model at all frequencies.

In addition, we compare the qualities of a Bessel-LPF-based reduced model and an ellipse-LPF-based reduced model in Figure 13.11. Figure 13.11(a) and (b) show the transient responses from the original circuit due to the square input waveform. Figure 13.11(c) and (d) show the transient responses from the Bessel-LPF-based reduced model while Figure 13.11(e) and (f) are the transient responses from the ellipse based reduced models using the same filter order. The results clearly show that the Bessel-LPF-based reduced model is superior to the ellipse-LPF-based based models. As shown in (Figure 13.11(d),(f)), the Bessel-LPF-based reduced model can effectively avoid overshoots and ringing. This result further demonstrates the rationale of the choice for a Bessel LPF over other types of LPF.

The second example is a RC circuit (168 nodes) with 132 terminals (14 drivers and 118 receivers). This circuit does not have to be reduced much owing to a large number of terminals compared with the number of nodes. But it serves as an example that the convex programming method fails to optimize because of the large terminal count. Still, we use the fitting method to perform the reduction in the frequency domain and make the reduced models accurate to 50 Ghz. We use a steep square waveform, as shown in Figure 13.12(a) as the input. We apply this signal to the original model, the reduced model, and the LPF-based reduced model. The output waveforms of these three models are shown in Figure 13.12(b),(c),(d), respectively.

By eliminating high-frequency components by LPF, the results are similar: output waveforms from the LPF-based reduced model (Figure 13.12(d)) match the original model (Figure 13.12(b)) well. But the simple reduced models lead to erratic time-domain behavior owing to their non-passivity at high frequencies, as shown in Figure 13.12(c).

The experimental results also show that the output of the LPF-based reduced model exhibits less ringing than the output of the original model. This is because the ringing is caused by high frequency components of the input signal and many of those components are eliminated by LPF in the reduced model. If the ringing is of interest, we can observe more of is by increasing the frequency range of the reduced models.

13.6 Summary

In this chapter, we presented a new passivity enforcement scheme for the passive modeling of general linear time-invariant systems. Instead of making the model passive for all frequencies as traditional methods did, the new method works on the signal waveforms, assuming that the circuit models are only passive for a limited frequency band (called conditionally passive). Such relaxation makes the circuit passive modeling work much easier using fitting-based methods for a reduced system, especially for systems with many terminals or requiring wideband accuracy using measured data. By shaping the signal waveforms such that the resulting signal's spectra are band limited, the resulting systems will still be passive from the simulation's perspective.

We presented the waveform shaping method using frequency truncation by means of FFT/IFFT and low-pass-filter-based approaches for transient waveform shaping processes. We analyzed the delay and distortion effects caused by using low-pass filters and the new methods to mitigate the distortion and delay effects. Numer-

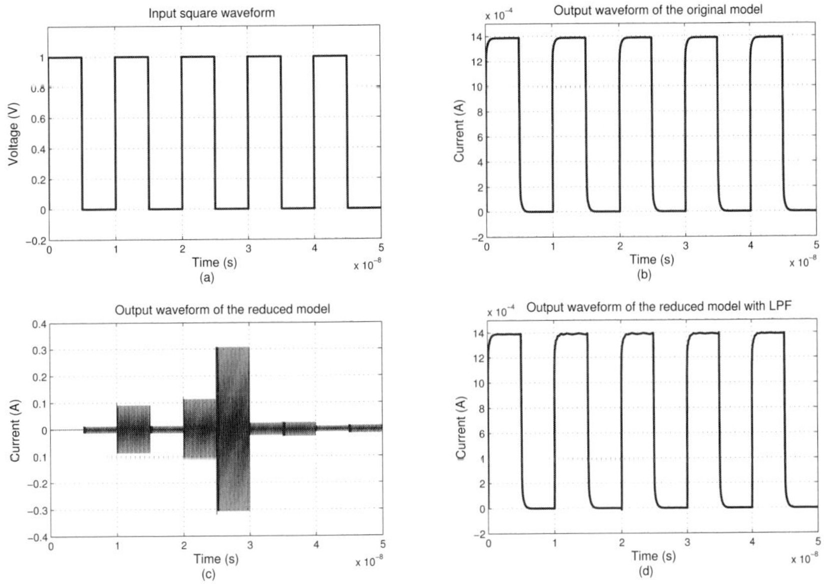

Figure 13.10 Comparison of responses of different models in time domain for the first example.

ical results on several interconnect circuits demonstrated the effectiveness of the method.

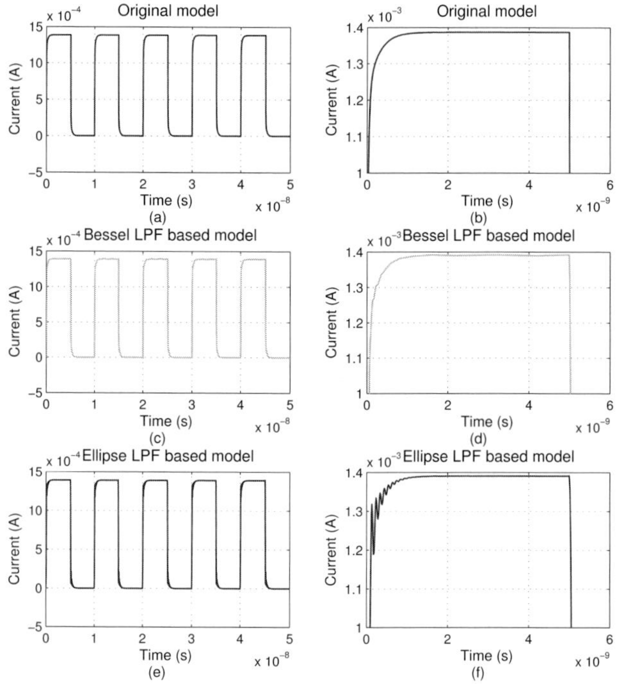

Figure 13.11 Comparison in time domain between reduced models based on Bessel filters and ellipse filters.

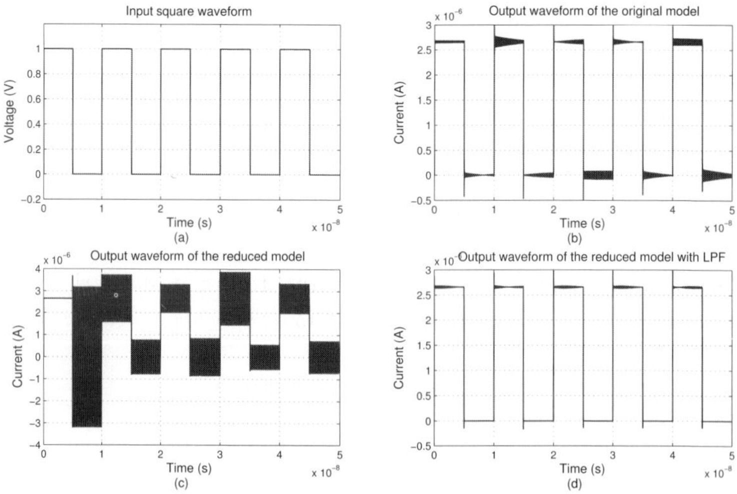

Figure 13.12 Comparison of responses of different models in time domain for second example.

References

1 R. Achar, P. K. Gunupudi, M. Nakhla, and E. Chiprout, "Passive interconnect reduction algorithm for distributed/measured networks," *IEEE Trans. on Circuits and Systems II: Analog and Digital Signal Processing*, **47**, (4), 287–301, April 2000.

2 D. Allstot, K. Choi, and J. Park, *Parasitic-Aware Optimization of CMOS RF circuits.* Kluwer Academic Publishers, 2003.

3 C. S. Amin, M. H. Chowdhury, and Y. I. Ismail, "Realizable RLCK circuit crunching," in *Proc. Design Automation Conf. (DAC)*, 2003, 226–231.

4 B. D. O. Anderson, "A system theory criterion for positive real matrices," *SIAM J. Contr.*, **5**, 171–182, 1967.

5 B. D. O. Anderson and S. Vongpanitlerd, *Network Analysis and Synthesis.* Englewood Cliffs, NJ: Prentice-Hall, 1973.

6 A. Antoulas and D. C. Sorensen, "Approximation of large-scale dynamical system: An overview," *Int. J. Appl. Math. Comput. Sci.*, **11**, (5), 1093–1121, 2001.

7 W. F. Arnold and A. J. Laub, "Generalized eigenproblem algorithms and software for algebraic Riccati equation," *Proc. IEEE*, **72**, 1764–1754, 1984.

8 W. E. Arnoldi, "The principle of minimized iteration in the solution of the matrix eigenvalue problem," *Quat. Appl. Math.*, **9**, 17–29, 1951.

9 M. Beattie and L. Pileggi, "Efficient inductance extraction via windowing," in *Proc. European Design and Test Conf. (DATE)*, 2001, 430–436.

10 P. Benner, E. S. Quintana-Orti, and G. Quintana-Orti, "State-space truncation methods for parallel model reduction of large-scale systems," *Parallel Comput.*, **29**, (11-12), 1701–1722, 2003.

11 S. Boyd, L. Vandenberghe, A. E. Gamal, and S. Yun, "Design of robust global power and ground networks," in *Proc. ACM Int. Symp. on Physical Design (ISPD)*, 2001

12 O. Brune, "Synthesis of a finite two-terminal network whose driving point impedance is a prescribed function of frequency," *Journal of Math. and Phys.*, **10**, 191–236, 1931.

13 R. Brune and R. Duffin, "Impedance synthesis without transformers," *J. of Appl. Phys.*, **20**, 816, 1949.

14 M. Celik, L. Pileggi, and A. Odabasioglu, *IC interconnect analysis.* Kluwer Academic Publishers, 2002.

15 C. Chen, *Linear System Theory and Design*, 3rd ed. Oxford University Press, 1999.

16 T. Chen, C. Luk, and C. Chen, "Inductwise: Inductance-wise interconnect simulator

and extractor," *IEEE Trans. on Computer-Aided Design of Integrated Circuits and Systems*, **22**, (7), 884–894, 2003.
17 T. Chen, C. Luk, H. Kim, and C. Chen, "INDUCTWISE: Inductance-wise interconnect simulator and extractor," in *Proc. Int. Conf. on Computer Aided Design (ICCAD)*, 2002.
18 E. Chiprout, "Fast flip-chip power grid analysis via locality and grid shells," in *Proc. Int. Conf. on Computer Aided Design (ICCAD)*, Nov. 2004, 485–488.
19 E. Chiprout and M. S. Nakhla, "Analysis of interconnect networks using complex frequency hopping," *IEEE Trans. on Computer-Aided Design of Integrated Circuits and Systems*, **14**, (2), 186–200, Feb. 1995.
20 E. Chiprout and N. S. Nakhla, *Asymptotic Waveform Evaluation.* Boston: Kluwer Academic Publishers, 1994.
21 C. P. Coelho, J. Phillips, and L. M. Silveira, "A convex programming approach for generating guaranteed passive approximations to tabulated frequency-data," *IEEE Trans. on Computer-Aided Design of Integrated Circuits and Systems*, **23**, (2), 293–301, Feb. 2004.
22 C. P. Coelho, J. R. Phillips, and L. M. Silveira, "A convex programming approach to positive real rational approximation," in *Proc. Int. Conf. on Computer Aided Design (ICCAD)*, Nov. 2001, 245–251.
23 J. Cong, L. He, C.-K. Koh, and P. Padden, "Performance optimization of VLSI interconnect layout," *Integration, the VLSI Journal*, **21**, (1&2), 1–94, Nov 1996.
24 W. J. Dally and J. W. Poulton, *Digital Systems Engineering.* Cambridge University Press, 1998.
25 A. Devgan, H. Ji, and W. Dai, "How to efficiently capture on-chip inductance effects: introducing a new circuit element K," in *Proc. Int. Conf. on Computer Aided Design (ICCAD)*, 2000, 150–155.
26 R. O. Duda, P. E. Hart, and D. G. Stork, *Pattern Classification*, 2nd ed. John Wiley and Sons Inc., 2001.
27 P. Elias and N. van der Meijs, "Including higher-order moments of RC interconnections in layout-to-circuit extraction," in *Proc. European Design and Test Conf. (DATE)*, 1996, 362–366.
28 W. C. Elmore, "The transient analysis of damped linear networks with particular regard to wideband amplifiers," *J. Appl. Phys.*, 1948.
29 D. F. Enns, "Model reduction with balanced realizations: an error bound and frequency weighted generalization," in *Proc. 23th IEEE Conf. Decision and Control*, 1984, 127–132.
30 K. K. et al., "Design of mixed-signal systems-on-a-chip," *IEEE Trans. on Computer-Aided Design of Integrated Circuits and Systems*, 1561–1571, 2000.
31 P. Feldmann, "Model order reduction techniques for linear systems with large numbers of terminals," in *Proc. European Design and Test Conf. (DATE)*, 2004, 944–947.
32 P. Feldmann and R. W. Freund, "Efficient linear circuit analysis by Pade approximation via the Lanczos process," *IEEE Trans. on Computer-Aided Design of Integrated Circuits and Systems*, **14**, (5), 639–649, May 1995.

33 ——, "Reduced-order modeling of large linear subcircuits via block lanczos algorithm," in *Proc. Design Automation Conf. (DAC)*, 1995, 376–380.

34 P. Feldmann and F. Liu, "Sparse and efficient reduced order modeling of linear subcircuits with large number of terminals," in *Proc. Int. Conf. on Computer Aided Design (ICCAD)*, 2004, 88–92.

35 R. Freund and P. Feldmann, "Reduced-order modeling of large linear passive multi-terminal circuits using matrix-Pade approximation," in *Proc. European Design and Test Conf. (DATE)*, 1998, 530–537.

36 R. W. Freund, "Passive reduced-order modeling of interconnects simulation and their computation by Krylov-subspace algorithm," in *Proc. Design Automation Conf. (DAC)*, 1999, 195–200.

37 ——, "SPRIM: structure-preserving reduced-order interconnect macromodeling," in *Proc. Int. Conf. on Computer Aided Design (ICCAD)*, 2004, 80–87.

38 R. W. Freund and P. Feldmann, "Reduced-order modeling of large linear subcircuits by means of the SyPVL algorithm," in *Proc. Int. Conf. on Computer Aided Design (ICCAD)*, 1996, 280–287.

39 ——, "Reduced-order modeling of large linear subcircuits by means of the sypvl algorithm," in *Proc. Int. Conf. on Computer Aided Design (ICCAD)*, 1996, 280–287.

40 F. R. Gantmacher, *The Theory of Matrices, vol. 1.* New York : Chelsea Publishing Co., 1977.

41 G. Gielen and R. Rutenbar, "Computer-aided design of analog and mixed-signal integrated circuits," *Proc. of IEEE*, **88**, (12), 703–717, Dec. 2000.

42 G. Gielen and W. Sansen, *Symbolic Analysis for Automated Design of Analog Integrated Circuits.* Kluwer Academic Publishers, 1991.

43 K. Glover, "All optimal Hankel-norm approximations of linear multivariable systems and their l^{∞}-error bounds," *Int. J. Contr.*, **39**, 1115–1193, 1984.

44 G. H. Golub and C. V. Loan, *Matrix Computations*, 3rd ed. The Johns Hopkins University Press, 1996.

45 E. J. Grimme, *Krylov projection methods for model reduction (Ph.D. Thesis).* University of Illinois at Urbana-Champaign, 1997.

46 S. Grivet-Talocia, I. A. Maio, and F. Canavero, "Recent advances in reduced-order modeling of complex interconnects," in *Proc. Dig. Electrical Performance Electronic Packaging*, Oct. 2001, 243–246, vol. 10.

47 G. Guardabassi and A. Sangiovanni-Vincentelli, "A two level algorithm for tearing," *IEEE Trans. on Circuits and Systems*, 783–791, 1976.

48 R. Gupta, B. Tutuianu, and L. Pileggi, "The Elmore delay as a bound for RC trees with generalized input signals," *IEEE Trans. on Computer-Aided Design of Integrated Circuits and Systems*, (1), 95–104, Jan. 1997.

49 B. Gustavsen and A. Semlyen, "Rational approximation of frequency domain responses by vector fitting," *IEEE Trans. on Power Delivery*, **14**, (3), 1052–1061, March 1999.

50 ——, "Enforcing passivity for admittance matrices approximated by rational functions," *IEEE Trans. on Power Systems*, **16**, (1), 97–104, Feb. 2001.

51 M. M. Hassoun and P. M. Lin, "A hierarchical network approach to symbolic analysis of large scale networks," *IEEE Trans. on Circuits and Systems I: Fundamental Theory and Applications*, **42**, (4), 201–211, April 1995.
52 Z. He, M. Celik, and L. Pillegi, "SPIE: Sparse partial inductance extraction," in *Proc. Design Automation Conf. (DAC)*, 1997, 137–140.
53 X. Huang, V. Rahjavan, and R. A. Rohrer, "Awesim: a program for efficient analysis of linear(ized) circuits," in *Proc. Int. Conf. on Computer Aided Design (ICCAD)*, 1990.
54 Y. Ismail, "Efficient model order reduction via multi-node moment matching," in *Proc. Int. Conf. on Computer Aided Design (ICCAD)*, Nov. 2002, 767–774.
55 Y. Ismail and E. G. Friedman, *On-Chip Inductance in High Speed Integrated Circuits.* Kluwer Academic Publishers, 2002.
56 ——, "DTT: direct truncation of the transfer function – an alternative to moment matching for tree structured interconnect," *IEEE Trans. on Computer-Aided Design of Integrated Circuits and Systems*, **21**, (2), 131–144, Feb. 2003.
57 J. D. Jackson, *Classical Electrodynamics.* John Wiley and Sons, 1975.
58 H. Ji, A. Devgan, and W. Dai, "Ksim: A stable and efficient RKC simulator for capturing on-chip inductance effect," in *Proc. Asia South Pacific Design Automation Conf. (ASPDAC)*, 2001, 379–384.
59 M. Kamon, M. Tsuk, and J. White, "FastHenry: a multipole-accelerated 3D inductance extraction program," *IEEE Trans. on Microwave Theory and Techniques*, 1750–1758, Sept. 1994.
60 M. Kamon, F. Wang, and J. White, "Generating nearly optimally compact models from Krylov-subspace based reduced-order models," *IEEE Trans. on Computer-Aided Design of Integrated Circuits and Systems*, **47**, (4), 239–248, 2000.
61 S. Kapur and D. Long, "IES3: A fast integral equation solver for efficient 3-dimentsional extraction," in *Proc. Int. Conf. on Computer Aided Design (ICCAD)*, 1997.
62 C. Kashyap and B. Krauter, "A realizable driving point model for on-chip interconnect with inductance," in *Proc. Design Automation Conf. (DAC)*, 2000.
63 K. J. Kerns and A. T. Yang, "Stable and efficient reduction of large, multiport RC network by pole analysis via congruence transformations," *IEEE Trans. on Computer-Aided Design of Integrated Circuits and Systems*, **16**, (7), 734–744, July 1998.
64 Y. Kim and D. Sylvester, "Indigo: http://vlsida.eecs.umich.edu/," *EECS Dept. at University Of Michigan.*
65 B. Krauter and L. Pileggi, "Generating sparse partial inductance matrices with guaranteed stability," in *Proc. Int. Conf. on Computer Aided Design (ICCAD)*, 1995, 45–52.
66 K. S. Kundert, *"A formal top-down design process for mixed-signal circuits," in Analog Circuit Design*, R. J. van de Plassche, J. Huigsing, and W. Sansen, Eds. Kluwer Academic Publishers, 2000.
67 C. Lanczos, "An iteration method for the solution of the eigenvalue problem of linear differential and integral operators," *J. Res. Bur. Standards*, **45**, 255–282,

1950.

68 A. J. Laub, M. T. Heath, C. C. Paige, and R. C. Ward, "Computation of system balancing transformations and other applications of simultaneous diagonalization algorithms," *IEEE Trans. Automat. Contr.*, **32**, 115–122, 1987.

69 S. Lee and M. H. Hayes, "Properties of the singular value decomposition for efficient data clustering," *IEEE Signal Process Letters*, **11**, (11), 862–866, Nov. 2004.

70 T. H. Lee, *The Design of CMOS Radio-Frequency Integrated Circuits*, 2nd ed. Cambridge University Press, 2004.

71 Y. Lee, Y. Cao, T. Chen, J. Wang, and C. Chen, "HiPRIME: Hierarchical and passivity preserved interconnect macromodeling engine for RLKC power delivery," *IEEE Trans. on Computer-Aided Design of Integrated Circuits and Systems*, **26**, (6), 797–806, 2005.

72 J. Lillis, C. Cheng, S. Lin, and N. Chang, *Interconnect analysis and synthesis.* John Wiley, 1999.

73 P. Liu, Z. Qi, A. Aviles, and S. X.-D. Tan, "A general method for multi-port active network reduction and realization," in *Proc. IEEE International Workshop on Behavioral Modeling and Simulation (BMAS)*, 2005, 7–12.

74 P. Liu, Z. Qi, H. Li, L. Jin, W. Wu, S. X.-D. Tan, and J. Yang, "Fast thermal simulation for architecture level dynamic thermal management," in *Proc. Int. Conf. on Computer Aided Design (ICCAD)*, Nov 2005, 639–644.

75 P. Liu, S. X.-D. Tan, H. Li, Z. Qi, J. Kong, B. McGaughy, and L. He, "An efficient method for terminal reduction of interconnect circuits considering delay variations," in *Proc. Int. Conf. on Computer Aided Design (ICCAD)*, 2005, 821–826.

76 P. Liu, S. X.-D. Tan, B. McGaughy, and L. Wu, "Compact reduced order modeling for multiple-port interconnects," in *Proc. Int. Symposium. on Quality Electronic Design (ISQED)*, 2006, 413–418.

77 R. Ludwig and P. Bretchko, *RF Circuit Design, Theory and Applications.* Prentice Hall, 2000.

78 T. Mangold and P. Russer, "Full-wave modeling and automatic equivalent-circuit generation of millimeter-wave planar and multilayer structures," *IEEE Trans. on Microwave Theory and Techniques*, **47**, (6), 851–858, June 1999.

79 Y. Massoud and J. White, "Simulation and modeling of the effect of substrate conductivity on coupling inductance and circuit crosstalk," *IEEE Trans. on Very Large Scale Integration (VLSI) Systems*, 2002.

80 K. Mayaram, D. C. Lee, S. Moinian, D. R. Rich, and J. Joychowdhury, "Computer-aided circuit analysis tools for RFIC simulation: algorithms, features, and limitations," *IEEE Trans. on Circuits and Systems II: analog and digital signal processing*, **47**, (4), 274–286, 2000.

81 B. Moore, "Principle component analysis in linear systems: Controllability, and observability, and model reduction," *IEEE Trans. Automat. Contr.*, **26**, (1), 17–32, 1981.

82 J. Morsey and A. C. Cangellaris, "PRIME: passive realization of interconnect models from measured data," *Electrical Performance of Electronic Packaging*, 47–50, Oct. 2001.

83 K. Narbos and J. White, "FastCap: a multipole accelerated 3D capacitance extraction program," *IEEE Trans. on Computer-Aided Design of Integrated Circuits and Systems*, **10**, (11), 1447–1459, 1991.

84 A. M. Niknejad and R. G. Meyer, "Analysis of eddy current losses over conductive substrates with applications to monolithic inductors and transformers," *IEEE Trans. on Microwave Theory and Techniques*, 166–76, Jan. 2001.

85 A. Odabasioglu, M. Celik, and L. Pileggi, "PRIMA: Passive reduced-order interconnect macromodeling algorithm," *IEEE Trans. on Computer-Aided Design of Integrated Circuits and Systems*, 645–654, 1998.

86 A. Pacelli, "A local circuit topology for inductive parasitics," in *Proc. Int. Conf. on Computer Aided Design (ICCAD)*, 2002, 208–214.

87 J. R. Phillips, L. Daniel, and L. M. Silveira, "Guaranteed passive balancing transformations for model order reduction," in *Proc. Design Automation Conf. (DAC)*, 2002, 52–57.

88 ——, "Guaranteed passive balanced transformation for model order reduction," *IEEE Trans. on Computer-Aided Design of Integrated Circuits and Systems*, **22**, (8), 1027–1041, 2003.

89 J. R. Phillips and L. M. Silveira, "Poor man's TBR: a simple model reduction scheme," in *Proc. European Design and Test Conf. (DATE)*, 2004, 938–943.

90 ——, "Poor man's TBR: a simple model reduction scheme," *IEEE Trans. on Computer-Aided Design of Integrated Circuits and Systems*, **24**, (1), 43– 55, 2005.

91 L. T. Pillage and R. A. Rohrer, "Asymptotic waveform evaluation for timing analysis," *IEEE Trans. on Computer-Aided Design of Integrated Circuits and Systems*, 352–366, April 1990.

92 L. T. Pillage, R. A. Rohrer, and C. Visweswariah, *Electronic Circuit and System Simulation Methods.* New York: McGraw-Hill, 1994.

93 Z. Qi, S. X.-D. Tan, H. Yu, L. He, and P. Liu, "Wideband modeling of RF/analog circuits via hierarchical multi-point model order reduction," in *Proc. Asia South Pacific Design Automation Conf. (ASPDAC)*, Jan. 2005, 224–229.

94 Z. Qi, H. Yu, P. Liu, S. X.-D. Tan, and L. He, "Wideband passive multi-port model order reduction and realization of RLCM circuits," *IEEE Trans. on Computer-Aided Design of Integrated Circuits and Systems*, 1496–1509, Aug. 2006.

95 H. Qian, S. Nassif, and S. Sapatnekar, "Power grid analysis using random walks," *IEEE Trans. on Computer-Aided Design of Integrated Circuits and Systems*, 1204–1224, 2005.

96 Z. Qin and C.-K. Cheng, "RCLK-VJ network reduction with Hurwitz polynomial approximation," in *Proc. Asia South Pacific Design Automation Conf. (ASPDAC)*, Jan. 2003, 283–291.

97 Z. Qin, S. X.-D. Tan, and C. Cheng, *Symbolic Analysis and Reduction of VLSI Circuits.* Boston, MA: Kluwer Academic Publishers, 2005.

98 Z. Qin and C. Cheng, "Realizable parasitic reduction using generalized Y-Δ transformation," in *Proc. Design Automation Conf. (DAC)*, 2003, 220–225.

99 B. Razavi, *RF Microelectronics.* Prentice Hall, 1998.

100 J. Rubinstein, P. Penfield, and M. A. Horowitz, "Signal delay in RC tree networks,"

IEEE Trans. on Computer-Aided Design of Integrated Circuits and Systems, 202–211, 1983.

101 A. E. Ruehli, "Equivalent circuits models for three dimensional multiconductor systems," *IEEE Trans. on Microwave Theory and Techniques*, 216–220, 1974.

102 Y. Saad, *Iterative methods for linear systems.* PWS publishing, 2000.

103 N. Sadegh, J. D. Finney, and B. S. Heck, "An explicit method for computing the positive real lemma matrices," *Int. J. Robust and Nonlinear Control*, **7**, 1057–1069, 1997.

104 B. Salimbahrami and B. Lohmann, "Krylov subspace methods in linear model order reduction: introduction and invariance properties," Institute of Automation, University of Bremen, Tech. Rep., Aug. 2002.

105 D. Saraswat, R. Achar, and M. Nakhla, "A fast algorithm and practical considerations for passive macromodeling of measured/simulated data," *IEEE Trans. on Advanced Packaging*, **27**, (1), 57–70, Feb. 2004.

106 A. Semlyen and A. Dabuleanu, "Fast and accurate switching transient calculations on transmission lines with ground return using recursive convolution," *IEEE Trans. Power Apparatus and Systems*, **94**, (2), 1975.

107 B. N. Sheehan, "TICER: Realizable reduction of extracted RC circuits," in *Proc. Int. Conf. on Computer Aided Design (ICCAD)*, 1999, 200–203.

108 ——, "Branch merge reduction of RLCM networks," in *Proc. Int. Conf. on Computer Aided Design (ICCAD)*, 2003, 658–664.

109 K. Shepard and Z. Tian, "Return-limited inductances: A practical approach to on-chip inductance extraction," *IEEE Trans. on Computer-Aided Design of Integrated Circuits and Systems*, **19**, (4), 425–436, 2000.

110 C.-J. Shi and X.-D. Tan, "Canonical symbolic analysis of large analog circuits with determinant decision diagrams," *IEEE Trans. on Computer-Aided Design of Integrated Circuits and Systems*, **19**, (1), 1–18, Jan. 2000.

111 ——, "Compact representation and efficient generation of s-expanded symbolic network functions for computer-aided analog circuit design," *IEEE Trans. on Computer-Aided Design of Integrated Circuits and Systems*, **20**, (7), 813–827, April 2001.

112 Y. Shi, H. Yu, and L. He, "SAMSON: a generalized second-order Arnoldi mehtod for multi-source network reduction," in *Proc. ACM Int. Symp. on Physical Design (ISPD)*, 2006.

113 M. Silveira, M. Kamon, I. Elfadel, and J. White, "A coordinate-transformed Arnoldi algorithm for generating guaranteed stable reduced-order models of RLC circuits," in *Proc. Int. Conf. on Computer Aided Design (ICCAD)*, 1996, 288–294.

114 M. Silveira, M. Kamon, and J. White, "Efficient reduced-order modeling of frequency-dependent coupling inductances associated with 3-D interconnect structures," in *Proc. Design Automation Conf. (DAC)*, June 1995, 376–380.

115 J. Singh and S. Sapatnekar, "Congestion-aware topology optimization of structured power/ground networks," *IEEE Trans. on Computer-Aided Design of Integrated Circuits and Systems*, 683–695, 2005.

116 R. Singh, *Signal integrity effects in custom IC and ASIC designs.* John Wiley and

Sons, 2003.

117 J. A. Starzky and A. Konczykowska, "Flowgraph analysis of large electronic networks," *IEEE Trans. on Circuits and Systems*, **33**, (3), 302–315, March 1986.

118 J. Sturm, "Using SuDuMi 1.02, a MATLAB toolbox for optimization over symmetric cones," *Optim. Meth. Softw.*, **10**, 625–653, 1999.

119 T. Stykel, "Gramian-based model order reduction for descriptor systems," *Math. Control Signals Systems*, **16**, 297–319, 2004.

120 H. Su, S. S. Sapatnekar, and S. R. Nassif, "Optimal decoupling capacitor sizing and placement for standard cell layout designs," *IEEE Trans. on Computer-Aided Design of Integrated Circuits and Systems*, **22**, (4), 428–436, April 2003.

121 S. X.-D. Tan, "A general s-domain hierarchical network reduction algorithm," in *Proc. Int. Conf. on Computer Aided Design (ICCAD)*, 2003, 650–657.

122 ——, "A general hierarchical circuit modeling and simulation algorithm," *IEEE Trans. on Computer-Aided Design of Integrated Circuits and Systems*, 418–434, April 2005.

123 S. X.-D. Tan, W. Guo, and Z. Qi, "Hierarchical approach to exact symbolic analysis of large analog circuits," in *Proc. Design Automation Conf. (DAC)*, June 2004, 860–863.

124 S. X.-D. Tan, Z. Qi, and H. Li, "Hierarchical modeling and simulation of large analog circuits," in *Proc. European Design and Test Conf. (DATE)*, Feb. 2004, 740–741.

125 S. X.-D. Tan and J. Yang, "Hurwitz stable model reduction for non-tree structured RLCK circuits," in *IEEE Int. System-on-Chip Conf. (SOC)*, 2003, 239–242.

126 X.-D. Tan and C.-J. Shi, "Hierarchical symbolic analysis of large analog circuits via determinant decision diagrams," *IEEE Trans. on Computer-Aided Design of Integrated Circuits and Systems*, **19**, (4), 401–412, April 2000.

127 G. C. Temes and J. Lapatra, *Introduction to circuit synthesis and design.* McGraw-Hill Book Company, 1977.

128 V. Valkenburg, *Linear Circuits.* Prentice Hall, 1982.

129 J. Vlach and K. Singhal, *Computer Methods for Circuit Analysis and Design.* New York, NY: Van Nostrand Reinhold, 1995.

130 J. M. Wang and T. V. Nguyen, "Extended Krylov subspace method for reduced order analysis of linear circuit with multiple sources," in *Proc. Design Automation Conf. (DAC)*, 2003, 247–252.

131 N. Wang and V. Balakrishnan, "Fast balanced stochastic truncation via a quadratic extension of the alternating direction implicit iteration," in *Proc. Int. Conf. on Computer Aided Design (ICCAD)*, 2005, 801–805.

132 J. White, F. Odeh, A. Sangiovanni-Vincentelli, and A. Ruehli, "Waveform relaxation: theory and practice," *Trans. Soc. Compu. Simul.*, 95–133, 1985.

133 W. Wolf, *Modern VLSI Design: System-on-Chip Design*, 3rd ed. Prentice Hall, 2002.

134 H. Yu and L. He, "Vector potential equivalent circuit based on PEEC inversion," in *Proc. Design Automation Conf. (DAC)*, 2003, 781–723.

135 ——, "A provably passive and cost-efficient model for inductive interconnects,"

IEEE Trans. on Computer-Aided Design of Integrated Circuits and Systems, **24**, (8), 1283–1294, 2005.

136 H. Yu, L. He, and S. X. D.Tan, "Block structure preserving model reduction for linear circuits with large numbers of ports," in *Proc. IEEE International Workshop on Behavioral Modeling and Simulation (BMAS)*, 2005, 1–6.

137 H. Yu, L. He, and S. X.-D. Tan, "Compact macro-modeling for on-chip RF passive components," in *Proc. IEEE International Conference on Communications, Circuits and Systems*, 2004, 199–202.

138 H. Yu, Y. Shi, and L. He, "Fast analysis of structured power grid by triangularization based structure preserving model order reduction," in *Proc. Design Automation Conf. (DAC)*, 2006, 205–210.

139 M. Zhao, R. V. Panda, S. S. Sapatnekar, and D. Blaauw, "Hierarchical analysis of power distribution networks," *IEEE Trans. on Computer-Aided Design of Integrated Circuits and Systems*, **21**, (2), 159–168, Feb. 2002.

140 G. Zhong, C. Koh, and K. Roy, "On-chip interconnect modeling by wire duplication," in *Proc. Int. Conf. on Computer Aided Design (ICCAD)*, 2002, 341–346.

141 ——, "On-chip interconnect modeling by wire duplication," *IEEE Trans. on Computer-Aided Design of Integrated Circuits and Systems*, **22**, (11), 1521–1532, 2003.

Index